John Hartig

We Are Not Alone

Civilizations
in
Outer Space

Book Cover
The Cosmic Cliffs Image
James Webb Space Telescope

Design and Writing
John Hartig
Second Edition © 2023

Published on Amazon and Kindle
with thanks for the chance
to publish affordably,
and the helpful phone chats

Dedication

<u>We Are Not Alone</u> is dedicated to the loving memories of my father, Michael Hartig, my mother, Rosa, my older sister, Nevenka, and my younger sister, Renate. They all have passed away.

In their time, they had no thoughts about outer space or space travel because they were concerned with daily things, like making a living and coping with illnesses.

In many ways, I was spoiled at university. My family gave me the opportunity to follow my own interests in literature, fantasy and science fiction. I am grateful to them for the doors that were opened for me.

John Hartig, author

Table of Contents

Foreword	p. 6
1 Interstellar	p. 9
2 First Contact	p. 12
3 The Find	p. 16
4 What to do?	p. 20
5 The Kardashev Scale	p. 23
6 Revised Kardashev	p. 27
7 What are the Chances?	p. 31
8 A Time Before Time	p. 37
9 The Fermi Paradox	p. 42
10 The Drake Equation	p. 45
11 Exoplanets	p. 49
12 New Discoveries	p. 54
13 TRAPPIST-1e	p. 57
14 The Watering Hole	p. 62
15 String Theory	p. 67
16 The Next Day	p. 71
17 The Day After	p. 75
18 Jon and Kate	p. 79
19 They Are Already Here!	p. 83
20 Another Day	p. 88

21 Unsettling Space Launch	p. 93
22 Meanwhile	p. 96
23 Next Step Stymied	p. 100
24 Incidental	p. 102
25 James Webb	p. 107
26 What's Your Address	p. 111
27 From Fiction to Fact	p. 114
The Keplerians	p. 116
28 What's in a Name?	p. 120
29 Parallel Injustice	p. 123
Aerth	p. 124
30 Scooped!	p. 127
31 Jon and Kathleen	p. 130
32 Erth	p. 133
33 Parallel Experiences	p. 137
34 What If?	p. 142
35 What you wish for	p. 147
36 The Space Trilogy	p. 151
37 What Forms?	p. 157
38 Meaning in the Universe	p. 160
39 God?	p. 163
40 The Gods	p. 168
41 A Swarm of Macrobes	p. 172
Resurgence of Hideous	p. 174
42 Kardashev and the Bible	p. 176

43 St. Peter Needs Help	p. 183
44 A Universe from Nothing	p. 189
45 Out of Nothing?	p. 194
46 Replacing Religion	p. 198
47 From Another Dimension	p. 202
The God Proton	p. 203
48 Rare Earth Hypothesis	p. 207
49 Video Interview	p. 211
50 The Future	p. 214
51 Conclusion	p. 219
Afterword	p. 227
Bibliography	p. 224
Acknowledgements	p. 225
Publication Contributions	p. 227
Books	p. 228
John's Photobook Series	p. 231
Contact	p. 234

Foreword

I make no excuse for my imagination when I mix facts with fiction. There are portions in my book, <u>We Are Not Alone</u>, where I fabricate names and whole civilizations, imagining what life might have been like on a prehistoric planet or on a technologically advanced world, way beyond us. Who knows, some of that might even be true!

SETI, the Search for Extraterrestrial Intelligence, is a non-profit institute that sends out broadcasts into space to say that humans are here. The world has already been transmitting high-frequency radio, television and radar for over 50 years. That is our "leakage" to tell the stars of our location!

There are people who object. Why tell aliens where we are? Astrophysicist, Michio Kaku, scoffs at this idea, noting that we have nothing to fear. If aliens are able to get to us from deep space, then they are advanced enough technology, that they don't need to plunder anything that we have on our little planet.

Yet, there are those who point out that Europeans colonized the lands of North America, only to decimate the tribes with disease and with stealing their lands. If Earth is colonized by an advanced civilization, then Earth could end up a slave planet.

These are things to think about as we reach out into space. Voyager I is 14.5 billion miles from Earth now, still moving forward in interstellar space. In about 40,000 years the spacecraft will drift within 1.6 light-years of a star in the constellation of Camelopardalis. An advanced civilization out there is not likely to investigate where this primitive relic came from. To them, Earth would not be worth the cost of a trip here, despite the imagination of authors like H.G. Wells in his book, <u>War of the Worlds</u>.

It is intriguing to think about alien life. Give me a cuddly ET anytime, who yearns simply, like he says, "ET Phone Home," rather than face the invasion of a belligerent species with tentacles. Sadly, before those scenarios happen, we might just obliterate ourselves anyway. Politicians are not the best representatives of our species.

Let's hope our interplanetary neighbours reach for the stars in terms of hope and inspiration, rather than for the greed that drives human nature, looking for gold and diamonds in outer space. Can you imagine what a mess a gold rush into outer space would make? Even though greed opened up the California of 1848 and the Yukon of 1898.

Colonization has to have a more civilized way of reaching out and expanding. Surely, we are more civilized than we were 100 years ago?

Let's hope we have become better than our pioneering forefathers. And that our interstellar neighbours, likewise, are better than the scary species in <u>War of the Worlds</u>.

*****

My goal in writing this book is to explain modern ideas about the Cosmos, about its Origins and about the possibility of Aliens. Hopefully, my writing style is simple enough to make complex ideas understandable for the average reader. We need to be educated by our astrophysicists and by our science fiction writers about the wonders that are "out there".

What was the Big Bang Theory? What is the Kardashev Scale? What is string theory? What is the Fermi Paradox?

Did God or a super being have anything to do with the whole of existence?

I've found YouTube videos instructive when physicists like Brian Greene, Brian Cox, Lawrence Krauss, Michio Kaku and David Kipping explain their ideas.

However, videos on outer space are scattered in an array of separate interviews over the internet. My book collates and gathers many of their speculations together in one book summarizing things in plain language.

I am pleased that the James Webb Space Telescope, hovering a million miles away from Earth, has been such a success. The photos and the telemetry, not hampered by space dust, will press the boundary of understanding about how our Cosmos formed. It's a great time to be alive.

> "Given the millions of billions of Earth-like planets, life elsewhere in the Universe without a doubt, does exist. In the vastness of the Universe, we are not alone."
> Albert Einstein.

1
Interstellar

Voyager I floated silently through interstellar space when its camera pointed backward and took a photograph of a distant Earth that looked like a pale blue dot approximately 3.7 billion miles away. The spacecraft, launched on September 5, 1977, kept heading further out toward the Andromeda Galaxy, after capturing its last farewell photo of how a tiny Earth, such an insignificant little dot in the Cosmos, looked on February 14, 1990. Voyager's circuits were still functioning.

How long would this man-made machine keep going? It showed how far mankind's reach could stretch into the twinkling cosmos. Voyager was now traveling at a percentage of the speed of light, having used the slingshot effect on several occasions as it swept past the gravitational fields of several large planets on its way into the gulf of interstellar space.

It set its eyes on Andromeda, an impossible 2.5 million light years away, an unattainable distance for an old relic like Voyager, except for the wormhole which had swallowed the unwitting craft, only to spit it out at the other end, close to Andromeda. Voyager reoriented itself and did what it was programmed to do in this new environment…to explore strange new worlds. It continued its silent trajectory and embraced a strange new destiny.

Voyager took many years to get this far before it took another snapshot of another pale blue dot, this time 4 billion miles ahead of it. It was a lovely little planet much like Earth. But this blue dot was not in our solar system.

Not even in our Galaxy. Andromeda had become Voyager's new home. The tiny blue dot, this time, was Irridon, a habitable world, unaware that humans had reached out from the Milky Way Galaxy to make contact with another world in another galaxy.

*****

Irridon, Type 1, a planetary civilization:

The Irridonians were a small, brown people. They looked very much like small monkeys; except they were hairless and bipedal.

The males of the species were 4.5 feet tall, and the females stood just over 4 feet. They were fine featured, quite nimble in hands and limbs. Their noses were flat, basically two holes above a fine mouth.

The species was mostly light brown, with basically no variations on the whole planet. No yellow people, no pink ones; no eyes that were slanted, although eye color and facial features varied. To humans they might all have looked alike. In their own eyes, they could distinguish variations of handsomeness among the men and beauty among the women.

Perhaps the sameness came from years of intermarriage. Perhaps, the small stature of their population was also due to the high gravitational pull of their planet, a planet which was a little larger than the Earth. Maybe, the differences were also an evolutionary thing, just the way Irridon took its course over the past 5 billion years. The planet had never seen a dinosaur.

Technologically, they were a very advanced species, far beyond what earthly technology was capable of. Instead of fossil fuels, they used fusion generators and harnessed the rays of their own sun. They sent their garbage into their sun with huge spaceships for incineration.

Anti-gravity hovercraft flew through their air spaces. The Irridonians had mined the ores of their moons long ago and now colonized many of their sister planets. They had developed specialties. Some planets grew plant-based foods; others, animal and protein-based foods.

The Irridonians never got sick. Many people lived to be 400 of our earthly cycles. Often, they became experts in several professions during their lifetimes.

The only barrier which they could not overcome was the speed of light, and obviously, the vast distances to other galaxies. They gave up searching for other life forms beyond their own galaxy long ago. They saw no benefit in discovering other life forms, subjugating them or plundering their lands and minerals. They had enough of their own raw materials.

Why contact a lesser life-form like Earthlings who could add nothing to the Irridonian lifestyle? At this point, they were not even aware of Earthlings.

The only reason that Voyager was now approaching Irridon was its accidental entrance through a wormhole in interstellar space. The alien spacecraft was detected by watchful astronomers, who were assigned by their government to keep an eye on errant asteroids which might do harm to their precious planet.

Irridonian astronomers had no idea that history was being made, that they had made first contact with alien life...all through a cosmic accident of a primitive Earth vessel slipping into a wormhole. Voyager 1 made contact!

2

First Contact

The Observatory stood proudly above the clouds, atop Mount Palo, on the planet Irridon, pointing its telescope upward to the Eastern horizon. At first, Voyager was mistaken for a tiny asteroid of no account; then, it took shape in the lens of the planet's most powerful telescope confirming that the object was indeed a fabricated thing coming straight toward the planet.

"It has a disk and an antenna, clearly made by some intelligent species," said Arrok, the senior astronomer of the Palo Observatory. "We must send one of our robotic ships to intercept it and bring it on board for study." This task was within the Irridonian scientific capability since robotic ships could achieve a good percentage of the speed of light, even though Voyager was still 5 billion miles away.

The funds and the approval for such an expedition was readily available since Irridon had been without wars for several Millenia. Its resources were focused on pure science now for the betterment of the entire planet. It operated on a one world government, with the purpose of finding more efficient ways to harness the energy from its own sun.

Arrok was eager to get his hands on the wayfaring spacecraft to see what kind of technology it had and to determine where the thing came from. He had the backing of GASA, Global Aeronautics and Space Administration. Arrok was proud that his son, Lyr, worked with him at the observatory. Like father like son!

It often happened that the offspring of scientific families would follow in the footsteps of their parents. It was a family tradition and only logical.

Both astronomers had detailed pictures now of what was approaching their home world since high-definition images had been sent back to their planet through their new long-range telescope, "The Irridon Eye", which had a mirror that was 100 times more powerful than anything Irridonian astronomers had previously possessed. The telescope floated beyond the asteroid belt which girded Irridon's solar system keeping an eye out on interstellar space like a protective guardian.

"It was worth it," said the father. "We now have clear images of what is approaching us. It's fascinating. I wonder where it came from?"

"It looked somewhat primitive," said the son, Lyr. "The antenna and general construction look like our designs from a thousand years ago. This is an old craft sent out by a primitive race, most likely that galaxy next to us some 2.5 million light years away. I wonder how it straddled that distance to get to us?"

"Probably through accident, not by design," said the father. "We have wormholes popping in and out of interstellar space every so often, and who knows what those entities swallow up and spit out?"

When the robotic ship intercepted Voyager, it had to anticipate the alien craft's trajectory and speed; slow down ahead of it and then speed up again to match the strange craft, so that the object could be brought on board by a tractor beam. Once the task was done, the robotic ship was instructed to head home so that the planet's scientists could get to work studying the captured alien craft.

*****

While that large drama unfolded among the distant stars, there was a romantic drama unfolding in the hearts of two Irridonians. Spica was the female astronomer on the team at the Palo Observatory. Lyr and Spica had eyes for each other.

The courting ritual for Irridonians certainly was unusual. The father, Arrok, had wondered two cycles ago why his son became absent-minded and why his work suffered, when it usually was so meticulously done. Pheromones were wreaking havoc between both astronomers whenever they were in close proximity to each other. Their mental faculties could not see straight.

Arrok wanted to get his son back to top efficiency. Otherwise, what good was he? Arrok finally advised his son: "Why don't you open up to Spica? Tell her how you feel. Go through the ritual for compatibility, and clear this matter up?"

Presided by Lyr's mother and father, Zeta and Arrok, and Spica's mother and father, Rigela and Ceros, arranged that the couple meet at an appointed time and place. The couple met, and in the presence of their parents, they tested their compatibility. They kissed and the kiss was sweet. They put their foreheads together and felt a telepathic joining, a union which showed that they were indeed suited to each other. This then was the official marriage ceremony.

Had the kiss not been sweet and the telepathic joining a failure, then the attraction between the two would have dissipated naturally like a passing fancy. Not all Irridonians got married. That was a joining for procreation and protection of their offspring. The Irridonians had ways to control their population. Most families had either one or two children. The total population of the planet was 5 billion, and every 10 cycles saw a huge exodus of extra people shipped off to colony planets. The ones which did not have a natural atmosphere for life had protective domes built which could house a whole city. The city was energized by the rays of the sun through solar panels.

Of course, there were many professionals who never married; they were simply married to their jobs.

Instead of wedding rings to show that they were not available, couples who married exuded a special hormone which basically said, not available, already taken. The Irridonians had lost the word for infidelity.

That week, friends and relatives were invited to celebrate Lyr and Spica's special union. Society loved to celebrate in a conservative, logical fashion, even though Earthlings might mistake them for being prudish.

3

The Find

It took almost one cycle for the robotic ship to bring Voyager home. Specialists opened the doors to the robotic ship and had a crane pull Voyager carefully out of the cargo bay.

Astronomers and technicians gathered around the alien craft in awe, deciding which way this interstellar intruder should be studied and possibly dissected. Everybody was instructed to look first and not to touch.

"One of my dreams come true," said Arrok to his son, as they stared at the disc and the gold covered body of the craft. Voyager lay prone in front of them, tilted to one side. Its disc, the high-gain antenna, had communicated with Earth for many years through the black void of space. It now faced an unresponsive ceiling in the Palo Observatory. The magnometer boom stretched out like the proboscis of a butterfly, a useless appendage. What could this advanced race of Irridonians learn from this cumbersome construct?

Lyr and Spica were less concerned about the mechanics of this unwieldy beast, than about the contents within various compartments which apparently had devices for data and communicating. Maybe, revealing the secret to where it came from and what kind of beings had launched it. The spacecraft lay there on its side, to all intents and purposes, dead.

"Let's hope our techs and our linguistic experts can unravel what the contents within this beast mean," said Lyr to his wife.

This was also Spica's dream come true, the reason she had begun her career in studying the stars. When she was a little girl, her parents took her camping.

"It will be good to get away from the city," they had said. "There you can see so many stars on a clear night."

Spica related this experience to Lyr early after their first meeting. He said that he had a similar experience from an early age, but much of that was due to his father's influence. Arrok's interest in the stars was infectious, and so Lyr's growing interest in astronomy led him to purchase a telescope. The investment was a natural extension of what the father passed down to his son.

"I read everything I could on stars and physics," said Lyr. "I knew what I wanted to do from an early age."

Both young people were enthralled by the wonder of the stars, their distance and their burning power.

"Maybe there is other life out there?" said Lyr one day to Spica. "We can always discover that together," she said, and a common bond blossomed into something stronger, and was happily solidified by their compatibility.

Now, a reality, proof of life beyond their own galaxy lay in front of them. What message did this thing to offer to them?

The unwieldy beast appeared to be covered in gold or silver foil. No, the cover was not gold at all, or even foil. It was a material which served a thermal function, as a multi-layer insulation. Technicians readily theorized that the shiny covering protected the craft from the coldness of space, and also served as heat protection against the radiation of a star.

The mechanics and techs uncovered the compartment in the spacecraft which held the Golden Record. But how do you read what's on it? Lyr and Spica knew it was some communicative device. The record contained greetings in

59 languages and one in the whale's language. There were also 116 pictures on Earth's cultures and science and 90 minutes of Earth's greatest musical hits, which included Mozart, Bach and Beethoven.

If the Irridonians had an old record player or a CD player, then maybe they could decipher what treasures lay inscribed in this Golden Record…but as it was, even they were befuddled by what the Golden Record could mean. Had the Earthlings had the foresight, they would have included a machine, a player, for the Golden Record, which was equipped with an automatic start mechanism and a visual screen. If you opened the lid that contained the player and its record, the thing would engage. Maybe nobody thought of such a setup on Earth when they put the communications system together.

Through their cleverness, the techs were able to finally figure out how the record functioned; they found a way to play it. They were astounded, not by the scientific achievement of this primitive race, but by the evident curiosity of those alien minds daring to send something out into space telling them all about themselves. How naïve was that!

*****

Arrok said, "If we were not such an advanced race, these people would be in trouble to reveal so much about themselves to the unknown cosmos."

Lyr and Spica took a different approach. Spica commented, "If this primitive race let us know that they existed, then wouldn't it be likely that there is another race out there who is above us, like we are above these Voyager people? Would it be wise for us to send out our own spacecraft to say, look at us, we are here!" Lyr and

Spica discussed the prudence of sending out messages into the unknown telling everybody that Irridon existed, where it was and images of what Irridonians looked like.

Lyr suggested that the revelation might not mean anything anyway. If there were other civilizations out there, then the distance of countless lightyears would be so great that actual contact would be impossible. Spica shrugged, "Distance is isolation and isolation is protection."

4
What to do?

The Astronomers Association of Irridon [AAI] had a sub-committee called SETI, the Search for Extraterrestrial Intelligence. It was a rather august group but small. Interest in its purpose had waned over the past 100 cycles. Some of its members felt that funding should be focused on the planet's own advancement, not in some futile search for life in outer space. Many were of the opinion that isolation was prudent. "What can aliens offer us, anyway?"

Scientists had deciphered the contents of the Golden Record and saw what Earthlings looked like, silhouettes of a man and woman, photos of everyday life on the planet, a traffic jam on the freeway. The sprinters in a race displayed a muscular Russian and behind him a black man trying to keep up. The Irridonians found that interesting, that these humans came in different colors.

"That could be a bone of contention," observed one scientist, "because differences always cause contention." The SETI group was not impressed by the demonstration of licking, eating and drinking by three Earthlings. "Barbaric," commented another scientist. What was more interesting to them was the photo which revealed that these Earthlings had X-ray technology, and also space travel, with an astronaut floating in his or her spacesuit in outer space.

The decision was finally reached by SETI. In this particular case, making contact with Earth was voted down. Earthlings were too primitive, and probably could not handle the knowledge of another civilization out there who were superior to them.

"We would gain nothing," said one committee member, "and most likely we would frighten the inhabitants of that planet."

Voyager I was moved into Irridon's museum which housed strange artifacts collected from outer space through the years. Therefore, nothing was ever sent back to Earth to let Earthlings know that intelligence existed beyond their "little blue dot".

"Most likely," said another member of the committee, "these humans also have religion. Letting them know about us, might create havoc in their belief systems and upset them. From our own long history, we know that leaders of such belief systems are jealous of their power. Who knows how we would unsettle their planet?"

Irridon had abandoned the concept of religion thousands of cycles ago and were happy to make their planet into a paradise. Inhabitants accepted their fate when they finally passed on at the age of 300 or 400. By that time, they had enough of life. Then, enough was enough and they were ready to embrace oblivion. They had no fanaticism about gods and the afterlife and were happy not to be passionate about any cause to the point that they'd die for it. No nationalism, no religious zealotry.

Arrok, Lyr and Spica accepted SETI's decision, and went on with their lives, studying the stars and how to harness the energy from their own sun for the betterment of all lives on their home-world.

They did, however, wonder if other lifeforms were out there who were either as sophisticated as they were or possibly advanced beyond the science of Irridon. Would those higher beings come to plunder our world, they asked? Or would they share their technology with us?

By the time Voyager made it into the museum, Lyr and Spica had two little children. They also had volunteered to settle in an off-world colony. The family dynasty of astronomers forgot about their quest for extraterrestrial life. They concentrated on their own galaxy. Irridon had a strong movement of individuals who were isolationistic.

"Let Andromeda take care of itself. Andromeda for Andromedonians."

Perhaps, there is something to a society becoming stagnant. And once it does, it regresses. Advancements were to be feared, now that contact was made with aliens, even though those aliens had inferior technology.

No acknowledgement ever came out of Irridon. Nothing was ever sent out to Earth to say, hello we are here, we are neighbours, we are friends. Irridon was a silent planet minding its own business within its own solar system. The contact with Voyager had only made the planet draw its horns inward, a strange reaction for a planet with such an advanced technology.

5
The Kardashev Scale

Type 0:

Earth had its own SETI or Search for Extraterrestrial Intelligence. SETI was incorporated as a non-profit corporation in 1984. Dr. Carl Sagan served on the Board of Trustees.

Carl Sagan wrote the book, <u>Pale Blue Dot</u>. He came up with the universal message which was sent out on the Voyager satellites. Carl Sagan was known as "the astronomer of the people" for popularizing scientific knowledge about outer space. He believed strongly that there was intelligent life on other planets.

> "When Earth dies some 5 billion years from now, there will be other worlds, stars and galaxies coming into being…and they will know nothing of a place called Earth…I would guess that the universe is filled with beings far more intelligent, far more advanced than we are." [<u>Pale Blue Dot</u>, 1994]

There have always been men of great imagination who have proposed life on other worlds, that we are not alone in the vastness of space. In fact, they have suggested that other beings on other planets are better and smarter than we are.

In 1964, twenty years before SETI was incorporated as a non-profit corporation, Nikolai Kardashev claimed that civilizations might exist on three levels, depending upon how they used energy. This has become known as the

"Kardashev Scale". Kardashev's Type I civilization was earthlike using fossil fuels on its own planet; Type II could tap into the energy of its own sun, and Type III could tap into the energy of its galaxy.

Since Kardashev's proposal, other astronomers have expanded the scale to reach a Type VII civilization and even Type XII.

Our Earthlike civilization has been demoted to Type 0, which has been redefined graciously to 0.72, bringing us close to Type I. Michio Kaku, physicist and futurist, suggested that humans may attain Type I status in 100 to 200 years. Then Type II in a few thousand years and Type III in 100,000 to a million years.

Carl Sagan suggested adding intermediate values to the Kardashev Scale, which would be produced by the formula:

$$K = \frac{\log_{10} P - 6}{10}$$

Where K is the Kardashev rating, and P is the energy it uses in watts.

Kardashev never defined Earth as a Type 0 which is quite humbling since we don't even count under a rearranged scale. But in 1973, we were given a number as a saving grace, 0.70, then in 2019, we snuck up to 0.73 on Sagan's redefined scale. However, this does not change the fact that we are still <u>Type 0</u>!

If a civilization controls its own weather, they are Type I. Anything planetary is Type I, as Michio Kaku says. We are not quite there yet.

Then if they can reach us from outer space, they are Type II. Type II has harnessed the energy from its star. They have developed "solar sails" which are thin constructs that are propelled by the radiation of a star. Michio Kaku says that Type II beings are immortal because they have conquered diseases.

Finally, Type III has mastered starships. They draw their energy from their galaxy. According to Michio Kaku, we have nothing to fear from advanced civilizations. "What do they want to plunder, if they have everything?"

*****

Michio Kaku was interviewed by Lex Fridman
YouTube: "Michio Kaku: We'll Make Contact with Aliens This Century"

Michio Kaku recalls an amusing incident in a confrontation with a kid after trying to explain Type 1, Type 2 and Type 3 civilizations, according to the Kardashev Scale.

A kid pestered the Professor, asking, "What about a Type 4 civilization?" "No," said Prof. Kaku, "the Kardashev Scale only allows for Type 3 civilizations!" "Well," insisted the kid, "What about the Continuum?"

Prof. Kaku thought about it and asked himself if the kid could be right. "Is there an extra source of energy, like the Continuum of Star Trek…the answer is Yes. There could be from dark energy…Dark energy is 73% of our universe …there could even be Type 5."

Now that is something to think about and it's scary!

*****

Astronomers have pointed out that the huge distance in inter-galactic space is the buffer that keeps us safe. Lightyears in interstellar space keep civilizations too far apart for contact, and therefore, civilizations remain silent among each other. Even travelling at 10% the speed of light, with theoretical fusion drives, it would take 30 million years to reach the Andromeda Galaxy from the Milky Way. Messages between the two would not be practical. Messages would easily get lost in time and distance.

It might be up to robots to colonize far away planets; it is estimated in as "little" as half a million years. Is it worth it?

6
Revised Kardashev Scale

It is estimated that the Milky Way, home to our Solar System, has 100 billion to 400 billion stars, and roughly one exoplanet per star in our galaxy. That makes at maximum a possible 400 billion exoplanets, a huge number of planets! Of course, you need to par that number down to habitable planets, with the right temperature, the right size and the right distance from its star. Still, a lot of planets!

Here is the list of a revised Kardashev Scale, which is a scientific way to also write the traditional notion of God out of the equation. It's amazing what science can come up with to close the gap between "the God of the gaps" and science.

- **TYPE 0,** basic *civilization*, Earth is at level 0.72 using fossil fuels like oil and natural gas. We are a baby civilization. It will take 100 years for us to squeeze into Type 1. At the advanced level of Type 0, we may use nuclear fission. We have nuclear reactors and the atomic bomb, but we are still too inefficient to jump to Type 1.
- **TYPE 1,** *planetary civilization*, which has nuclear power as its main power source. Nuclear fusion is used, capable of living in the sea and in the clouds. Able to harness the power of its sun.
- **TYPE 2,** *stellar civilization*, no poverty, uses Dyson sphere to harness energy from its star. A Dyson

sphere is a mechanical construct built around a star to harness huge amounts of power. They have developed "solar sails" which are thin constructs that are propelled by the radiation of a star. Michio Kaku says that Type II beings are immortal because they have conquered diseases. We become type 2 in a million years.

- **TYPE 3,** *galactic civilization*, gathers energy from the galaxy. They use starships, solar powered. Can use wormholes for transportation. They have a directive to remain silent and not interfere with primitive civilizations. Interstellar space is so distant, that they would not want to spend effort to come to us. We have nothing to offer them. Life span is extended.
- **TYPE 4,** *universal civilization*. Create their own wormholes. Uses Supernovas for energy. Can travel to other galaxies to gather energy. They are able to transcend dependence upon physical bodies. Type 4 life forms might have seeded life on habitable planets and might have appeared to us as gods. Not omniscient though. They realize that multiple universes are true.
- **TYPE 5,** *multiversal civilization*, able to access white hole energy, have access to other universes. Lifespan is extended.
- **TYPE 6,** *multidimensional civilization*, realizes that life exists beyond the third dimension, discarded their bodies a long time ago and exist as pure energy forms, could travel backward and forward through time. They exist in an infinite number of multiverses that represent an infinite number of instances and all laws of physics. They are capable of creating and maintaining the laws of physics.

Would realize that a type 7 civilization exists but would not attain it.
- **TYPE 7,** *creator civilization*, is not a civilization, as such. It is existence itself. They transcend the Omniverse which is the collection of reality, all universes. They would not be gods, however. This speculation would upset religious beliefs of lower forms and cause great controversy.

*****

Leave it to astronomers and physicists to redefine and add more divisions to the Kardashev Scale. But that's the nature of the beast, to see things where no man has seen things before!

Physicists see 30 and 50 types of civilizations, and even an Omega civilization where existence resides within computer simulations. Type 40 is a civilization that controls at least one "ultrabox"! Those beings make their own laws of physics.

For the purpose of this book, however, it would be cumbersome to talk about 30, 50 or even an Omega civilization.

As to the question of how many civilizations there are within Type 0 on Earth, sociologists have subdivided Type 0 into 36 subdivisions.

This refinement is not the Kardashev Scale. These 36 are subsumed under Type 0. To subdivide civilizations further is confusing, even though scientists love to see fine distinctions and come up with more crazy insights.

But for this book, let's stick to the Kardashev Scale as defined by energy consumption and how outer space civilizations might use energy and be categorized in a hierarchy of advancement.

Our Earth is used as a standard for how life might get started, which requires a rocky planet in an appropriate distance away from a suitable star, after about 5 billion years of time. This poses the possibility that there could be other "us"-es out there at different levels of development.

7
What Are the Chances?

The urge to find beings in outer space like us is a strong urge. Maybe, it comes from something primitive in us.

Astrophysicist Sara Seager confessed in her book, <u>The Smallest Light in the Universe</u>, that she was almost ready to change her career, to give it all up to become a veterinarian. Her father had always wanted her to become a physician like him. When *51 Pegasi b* was found in October 1995, she decided to stick it out and become an astronomer to study the stars.

Sara Seager entered a whole new world of exploration on exoplanets through her basic urge to find out what was out there.

> "The sun couldn't be the only star that had accumulated planets. But proof of their existence, never mind their potential inhabitants, remained...out of reach." [<u>Smallest Light</u>..., Seager]

The problem was, of course, distance. A trip across the Milky Way could take almost 2 billion years. The Milky Way was one of hundreds of billions of galaxies! Even a trip to our nearest star, Alpha Centauri, could take 50 million years! We'd be looking at hibernation either to wake up to a dream or a nightmare. Better to just send self-replicating robots into the void.

As of yet, we have not taken photos of the actual surfaces of exoplanets. We guess that exoplanets are out there by a tug effect on their sun, a gravitational effect called "a wobble". Also, there's a blinking out effect as the exoplanet dims the light of its sun during its orbit. That's called "a transit".

Look at the math, say scientists who study exoplanetary systems. More than a billion trillion sunlike stars exist in the observable universe. A quarter of all stars are orbited by planets of Earth's size and surface temperatures, planets that have liquid water on them! In the Milky Way alone, there are 50 billion earthlike planets. [Extraterrestrial, the first sign...Avi Loeb, 2021]

Carl Sagan guesstimates that the number of habitable planets with extraterrestrial intelligence might be 1 billion. Science fiction author and biochemist, Isaac Asimov, through a principle of mediocrity, rounds that figure down to 390,000 civilizations capable of technology. Even those are high numbers!

Carl Sagan figures that there are 400 billion stars in the Milky Way with innumerable planetary systems. Maybe civilizations arise all the time but wipe themselves out as soon as they are able. Look at the Earth with its capability of nuclear weapons and the continued threats of war among its own peoples.

The race to space has not been pure or unsullied. Wernher von Braun developed the V2 rockets which he lovingly thought might someday aim at the Moon, but instead aimed at London, England, with Hitler's master plan to win World War II. Von Braun was made an SS Officer by Hitler for his work of mass destruction. The United States forgave him his sins, so that he could divert his inventive mind to advance the U.S.A. in its space race against the Russians.

There are many people who say, what is the sense in aiming at the stars, when we have our own problems here on Earth to solve, such as healthcare, infrastructure and defense. Is it really worth it to mine petroleum on Titan and transport it back to Earth? Or to mine diamonds and gold in the depths of Mars?

*****

Yet, human curiosity and imagination are difficult to suppress. We always look up and ask ourselves, how do we get there?

The Ancient Greeks were a smart people. It is amazing what they figured out, centuries before Europe sank into its Dark Ages and then into its Medieval Age with the Catholic Church using the Inquisition to get rebellious people to get back to "right" thinking.

Anyway, let's look at some highlights of Greek innovative thought before Christianity took a stranglehold on the free mind:

- Democritus, in about 400 B.C., proposed that matter was made up of small things called "atomos" which means uncuttable.
- Then, about 240 B.C., Eratosthenes used geometry to figure out the circumference of the Earth to be about 250,000 stades or 28,000 miles. The Greeks already had a notion that the Earth was a sphere revolving around the sun.
- Aristotle believed that the Earth was a sphere in the 4th century B.C. He mentions Eudoxus who regarded each celestial body as a set of spheres, nested one inside the other.
- Ptolemy published the idea in the <u>Almagest</u> in the second century A.D. He came up with the geocentric idea of an Earth centered solar system with a sun revolving around it.

- Around 165 A.D., a Greek writer, Lucian of Samosata, wrote an imagined account of a trip to the Moon. The Moon beings and the Sun beings were human in appearance and politics. They argued about colonizing Jupiter!

*****

After all the great insights from the Greeks, it is surprising that Europeans regressed through the Dark Ages, [from about 476 C.E. to about 1000 C.E] and even to the time of Galileo in 1610.

Europeans reverted and embraced ignorant thinking, stifled by the dismal blinders and threats of the Inquisition. The Catholic Church made sure that Christians in Europe held to "right thinking" even into the Renaissance! And yet, even amidst such censure, creative minds searched the Cosmos for new understanding because they had that urge, "to be curious".

In the 15th century, Nicholas of Cusa, a Cardinal, wrote that space was infinite, that there was no center to the universe, that all things moved, including the Earth. Stars were distant suns that were attended by planets that were inhabited.

He wrote a treatise where he imagined a conference in Heaven where representatives of all religions came to agree on one religion which would ensure Peace of Earth. He did, however, order Jews of Arnhem to wear badges to identify them as Jews because he thought they were too stubborn to concede to a unified religion. Nicholas died in 1464, luckily protected by the Pope, and before the Inquisition accused him of heresy.

In 1543, Nicholaus Copernicus proposed a radical theory of the universe, a heliocentric solar system, where the Earth was only one planet among others which orbited the Sun. Copernicus is the father of modern astronomy. His idea took more than a century to become accepted. Fifty years later, Galileo proved the Copernican notion was true through the use of his telescope.

In 1609, Galileo discovered that there were craters on the moon by use of his telescope. He also discovered that Jupiter had moons; he saw the rings of Saturn, and that there were many more stars than one can see with the naked eye.

For his heliocentric assertions, he was put under house arrest by the Catholic Inquisition. He insisted that his accusers look through his telescope for themselves. They refused. The Bible contradicted his ideas. The thought police of the Catholic Church refused to see beyond their narrow beliefs, and thus, hampered the advancement of science.

Isaac Newton published the *Principia Mathematica* in 1687 in which he established classical mechanics. He is famous for the three laws of motion which form the basic principles of modern physics. He also studied the effect of white light through prisms. He along with Leibniz discovered calculus.

Charles Darwin published *On the Origin of Species* in 1859. He explained the idea of evolution and natural selection, and of course, dispelled the notion that the world was only 6,000 years old. Fundamental Christians were outraged by such unorthodox thinking.

In 1897, H.G. Wells wrote *War of the Worlds*, where Martians declared interplanetary warfare on Earth. The Martians were eventually defeated by a Pandemic of bacteria.

The history of scientific discoveries and religious beliefs is scary. To debunk ancient beliefs, inventive thinkers could get excommunicated and burned at the stake in the 13th century.

Now, science has unraveled so much to dispel old ignorance. Yet, as science explains more, "the God of the Gaps" gets smaller and smaller. It is difficult for religious people to tear themselves away from fundamentalism. They do not see that their God may just be bigger than they think! They need to have a wider and richer vision of their God, "if He holds the whole world in his hands."

Albert Einstein published his "Theory of Special Relativity" in 1905 and his "Theory of General Relativity" in 1915. Theories of relativity superseded Newtonian mechanics.

In 1905, Einstein claimed $E=mc^2$. This said that energy and mass are interchangeable. If the atoms in a paperclip could release their energy, it could destroy a whole city. He also claimed that time and space are interchangeable, and that they should be thought of as one unit, "spacetime". General relativity predicted that gravitational waves would bend light as it travelled past a star. Indeed, this was proven in 1919 during a solar eclipse.

Father Georges Lemaitre, a Belgian priest, formulated the notion of "the big bang theory", though he did not use the term to describe it. He claimed that the universe began in a cataclysmic explosion from a primeval super-atom. Ironically, Einstein said to Lemaitre, "Your calculations are correct, but your grasp of physics is abominable." It was later in 1949, when English astronomer, Fred Hoyle, actually coined the term, Big Bang, on a BBC radio broadcast, intending the term to make fun of the idea.

8
A Time Before Time

Type 0 primitive:

Grokk came back to the cave with the kill for the day. He had already eviscerated the beast outdoors. When he felled the gazelle with his wooden spear, the tip hardened by fire, he knelt beside the bleeding creature and gave thanks to the Great Spirit for this gift of food and life, and also the fire.

He grunted and waved for his tribe to come and help butcher the animal. They would have plenty to eat for the next week.

His mate, Shema, was proud of him. A strong warrior, a provider. He was 5 feet tall, covered with black hair, so much like the ancestors from which he and his tribe sprang. His nose was flat with two nostrils just above his thick upper lip. His eyebrows were pronounced, reminiscent of the ancient people he came from.

Some of his ancestors still lived in the old way, in the trees and refused to come down to walk erect and proud on two legs on the ground. He and his tribe were different. They were aware of their own existence and their mortality. They had the use of tools, and they were aware of a future that they had to plan for. They knew winter was coming soon.

They had the use of fire, which they could start with sticks and flint. Other tribes thought the ability to start a fire was magic and refused to learn how. Grokk tried to show them how this magic worked, but they shunned him and threatened to stone him. He tried to grunt an explanation, in his own way, that meat cooked in fire tasted so much better than raw flesh. If only they would understand and learn!

Nothing removed the dark curtain of ignorance. Grokk went his own way. He soon had a helpmate, Shema, who supported him and worked with him. The tribe made Grokk chief, since he was the wisest among them, and the strongest. Grokk accepted a crown of laurel leaves, suggested by the tribe's medicine man. There was a ceremony which recognized him as king. His wife became their queen.

Grokk did not know what to do when another tribe infringed on his territory. They were jealous of his prosperity; how he always killed game which eluded them. They declared war. He prayed to the Great Spirit. If peace was not possible, how could his tribe conquer the threat?

While Grokk was walking past a bush one day with his spear and a stick in hand, a branch caught the jagged edge of his stick. He pulled against it and his little stick shot forward out of his hand with great momentum. Grokk was amused by this. He tried it again, and an inventive idea formed in his head. What if he cut a solid branch, about his height, and tied each end with a thin strip of leather, making sure there was a good curve to his branch? He could fling thin sticks quite a distance if he pulled it back with the twine and let go. It worked. He sharpened the stick and hunted small game with the new weapon. Eventually, he fashioned a fletching out of a feather, and then a pointy flint arrowhead with which he could shoot bigger game.

Grokk taught the use of this weapon to the other men in his tribe. They had plenty to eat.

There is a primitive instinct in all creatures to mark their territory and to say, "I was here". Grokk painted crude pictures of gazelle on the walls of his cave, showing a hunt scene with spears, bows and arrows. He then made his mark by taking a mouthful of liquified ochre and spraying that upon the hand he placed on the cave's wall. It said not only Grokk was here but also that Grokk drew this. Grokk was proud of his artwork.

While Grokk's men were out hunting one day, the other tribe waylaid one of his hunters and seized his bow and arrow. They did not know how it worked. They stoned the warrior who carried this magical weapon.

Grokk took this as a challenge to go to war. Either they join his tribe, or there would be bloodshed.

Grokk climbed a cliff to get a high vantage point from which he could observe the movements of his enemy down below.

At the top of the plateau, he sat there momentarily mesmerized by a little ant near him which was moving a dried leaf, at least 10 times its size. The ant dropped the leaf and zig-zagged aimlessly in circles as if disoriented. It picked up the leaf again, moved it again, dropped it, scurried in circles once more, came back to pick up the leaf, only to move it once more and drop it in an aimless ritual. Perhaps, the Great Spirit sees us like that?

Did the ant know what it was doing? What was the purpose of the ant's repeated indecisive ritual? The ant's actions seemed futile to Grokk.

He peered over the plateau where he could spy on his enemies who were moving in and out of their cave. Maybe, the ant was like them, performing rituals with no apparent purpose.

Grokk accidentally dislodged a rock on the opposite side of the plateau and watched it bounce down the cliff where it dislodged another rock, a big boulder shaped like a flat disk. It rolled down the hillside for a long way, maintaining a vertical balance. It finally fell sideways to fall over and come to rest.

Grokk was lucky that no one noticed the noise from the opposite side of the cliff. The incident, however, gave Grokk an idea which he would investigate later. There was always tomorrow.

If he could only get 4 large flat circles made of wood, put holes in their center and somehow join them with a long pole. Maybe, put a platform on top? Maybe you could pull things on top of this contraption. He would investigate the idea but on some other day. There was more pressing action to be taken at the moment.

Grokk came down from the cliff, gathered his men together. They were armed with spears and bows and arrows. They attacked and defeated the other tribe easily, leaving only the women and children alive. These they assimilated into their own tribe, making his tribe larger and stronger.

Grokk demanded their obedience, and peace was agreed upon. Grokk thanked the Great Spirit for his victory. The larger tribe prospered, and they were feared throughout the land. There was plenty of food for all of them. Grokk bowed his head to the Great Spirit which lived in the Sun and gave his world warmth and life.

Peace reigned in the kingdom.

One day, Grokk saw something strange in the sky which was unsettling. A huge round orb crossed its shadow over the sun blocking out the light of life. The sun suddenly exploded, and all existence immediately incinerated. It was a swift obliteration. Grokk and all his people went the way of their ancestors. Life ceased to exist in Grokk's time and space.

Grokk's world existed some 18 billion years ago, before the Big Bang of our own universe.

Sadly, Grokk's universe met its untimely end when a rogue planet ran into his Sun, the very Sun he worshipped for life. This set off a chain reaction, so that all existence shrunk into a singularity, a black hole, the size of the head of a pin.

In a time before time, that singularity exploded in a Big Bang so that a new universe came into being, ours, giving birth to new stars, and new planets. The Milky Way and other galaxies formed and were thrown outward in this new reality. Our Earth was an offspring from Grokk's dying universe.

With 400 billion stars in the Milky Way, it is estimated that there are more than 1 <u>trillion</u> orbiting planets, and many of those planets, it is estimated, hold the potential for life.

Astrophysicists say that our universe is 13.8 billion years old. The observable universe [including dark matter and dark energy] is ever expanding, ever running away from us, without a thought that Grokk ever existed, nor his world and his timeline which vanished so long ago.

9
The Fermi Paradox
1950

Enrico Fermi, an Italian physicist, was known for his work on the first nuclear reactor. He has made contributions to quantum theory and nuclear particle physics.

He asked the question in 1950, if there is extraterrestrial life out there, where is everybody? If there are billions of planets capable of supporting life, then the law of large numbers indicates that extraterrestrial life should not be rare! Why haven't any visited Earth?

His question has led to some interesting speculations:

- **Down the Rabbit Hole:** We have not found evidence of extraterrestrials because our reality is an illusion. As in, we are living in a computer simulation created by our alien overlords, who make the rules.
- **Our Wires are Crossed:** It's very possible that other intelligent life forms are actively sending transmissions into space. We simply don't use the same range of frequency radio waves or perhaps everyone is listening and no one's talking.
- **Earth is a Fishbowl:** In this scenario, alien civilizations know that we exist. They simply watch our development from afar to let us evolve without influence, ultimately forging our own path.

- **Destroy or be Destroyed:** In Darwin's theory of evolution, a tenet says that the strongest species survives. A similar tactic may be employed by alien beings; perhaps they stay silent, hoping that hostile species aren't alerted, or they strike before others destroy them first.
- **Space is Too Vast:** Space, simply put, is incredibly large. So large, if we beamed a transmission to the closest star, it would take 4 years to reach the system. Say intelligent life did pick up one of our signals, it might take years, if not decades, to get a response.
- **They are Already Here:** It would be silly to presume that all life is similar to Earth's. Perhaps alien beings are so different, they would not register to us, even if they were under our noses. Conversely, they may be so similar, they are indistinguishable from humans, and can readily avoid detection.
- **They Live in Unlikely Places:** We don't know where to look. The search for life is largely conducted on other planets but what if we are looking in the wrong place altogether! A truly advanced civilization may not be anchored to a rocky world. In fact, some astronomers suggest that, because of energy demands, aliens might lurk on the edge of the galaxy, maybe even in supermassive black holes themselves.
- **They use Tech to Spy:** Regardless of how technologically advanced a civilization becomes, space exploration will always be long and fraught with danger. Instead of sending manned ships to explore the galaxy, aliens might dispatch self-

replicating nanobots, like "von Neumann probes" to do the work for them.
- **We are the Aliens:** Perhaps eons ago, some alien race visited Earth. After seeing all the earmarks of a habitable world, the creatures sowed the seeds of life with their own genetic material, before going along their merry way. We, in a sense, are their experiment.
- **Life is Extremely Rare:** Perhaps, in the search for extraterrestrial intelligence, the simplest solution is the correct one. We haven't encountered signs of life, either because it doesn't exist, or it's exceedingly rare. The prerequisites for complex life are nearly impossible to replicate in their entirety elsewhere.

*****

We must consider "abiogenesis" in these scenarios. This is the original idea of life or living organisms evolving from inorganic or inanimate substances, i.e., a form of spontaneous generation. To construct any convincing theory of abiogenesis, we must take into account the condition of the Earth about 4 billion years ago. The usual theory involves a "Primordial Soup."

10
The Drake Equation
1960

Genesis: "and God brought Abraham forth abroad, and said, look now toward heaven, and tell the stars, if thou be able to number them, and God said unto Abraham, so shall thy seed be."

The number of stars that can be seen with the unaided eye is about 6,000 but taking the atmosphere into account probably 2,500. However, any way you look at it, it was a big promise God made to Abraham about his hegemony on Earth.

Science now tells us that, the greater number of stars, the greater the chance of numerous life forms.

The number of stars in the Milky Way Galaxy is estimated to be 400 billion! The number of habitable planets that orbit these billions of stars would also be astronomical, maybe a trillion planets! That is a large possibility for holding life! Do the math! What are the chances that some forms of life have developed on some of these other planets?

In fact, somebody already did the math. In 1960, Dr. Frank Drake came up with "the Drake Equation".

He did this in preparation for a meeting in Green Bank, West Virginia. Drake is considered the father of SETI [the organization for the Search for Extraterrestrial Intelligence]. In simple terms, the Drake Equation is a probabilistic argument used to estimate the number of possible active, communicative extraterrestrial civilizations in the Milky Way.

The Drake Equation reads:

$$N = R_* \times f_p \times n_e \times f_l \times f_i \times f_c \times L$$

The various terms are defined as follow:
- **N** = the number of currently active, communicative civilizations in our galaxy.
- **R**∗ = the rate at which stars form in our galaxy.
- fp is the fraction of stars with planets.
- **ne** = the number of planets that can potentially host life, per star that has planets.
- **fl** = the fraction of the above that actually do develop life of any kind.
- **fi** = the fraction of the above that develop intelligent life.
- **fc** = the fraction of the above that develops the capacity for interstellar communication.
- **L** = the length of time that such communicative civilizations are active. Note that "fraction of the above" means that all the previous conditions have been satisfied. For example, when we consider fc we assume that intelligent life has already developed.

*****

We are 26,000 lightyears from the center of the galaxy. Can you imagine taking 26,000 years to get there at the unimaginable speed of light!

At the low end, we estimate 200 billion stars in the Milky Way, with an upper estimate of 400 billion stars. It becomes unwieldy to scour through a trillion exoplanets to see if there is life on them. Astronomers are using machine learning algorithms to search for patterns of communication in the radio spectrum.

We use radio signals because it can pass through interstellar dust. We are also using infrared radiation to see through that dust.

Lasers have been suggested in recent years as a means of signaling outer space. Professor Cooper pointed out that compact computer chips could be beamed into interstellar space on laser beams. At fifth of the speed of light, they would reach Alpha Centauri in 20 years, quite reachable in a human lifetime.

Professor Cooper mentioned the "contact paradox", where we want to contact aliens badly, but once done, there could be unwanted consequences. These aliens may be 1,000 years in advance of us. Some scientists fear that if aliens colonize us, we may end up victimized like the Aztecs and Incas were when the Spanish conquistadors conquered North America in the 16th century.

We should do a little reconnaissance first. A prediction by astronomer, Keith Cooper, says that by the end of this century, if there is life out there within 50 lightyears of us, we will find it!

There are other possibilities when you think about alien life on other planets.

During the universe's 13.8-billion-year existence, life on other planets could have come and gone already, billions of years before life ever formed on Earth. When mankind ceases to exist on Earth and becomes extinct, who knows, if another planet may fill the gap to replace us, and then once that life-form itself becomes extinct, what other life-form may spring up in its place perhaps in a galaxy far, far away, not even our own!

The universe is so large; time is so vast; stars and planets come and go. We must not think that we are God's sole creation in the universe.

Physicist, Michio Kaku, observed that the laws of Einstein break down at what you call, "the Planck Energy" level. He sees the Planck Energy level applied to string theory which demands other dimensions. Professor Kaku commented that any civilization that harnesses the Planck Energy becomes masters of time and space. Any civilization at that level would consider Earth not even worthwhile invading or even thinking about.

11
Exoplanets

Type 0 advanced, that's us!

"Outer space" is the physical universe beyond Earth's atmosphere. The vacuum of space is really not a vacuum. Starshine in the form of photon packets travels through the vacuum of space. There is also dust, mini-meteorites, asteroids and quantum particles zipping around. It is a chaotic and dangerous place out there.

The James Webb Space Telescope was hit by a micro meteorite sometime between May 23 and May 25th, 2022. Astronomers say that despite the damage, the machine is still functional and will give us images and data beyond what Hubble is able to do. Scientists hope they will yet make the 10-billion-dollar investment in the machine worth our while.

The James Webb Space Telescope [JWST] is orbiting our sun 1 million miles away from Earth. Voyager I is floating out there too somewhere in interstellar space at 14.5 billion miles away from Earth. We are a pale blue dot to it. The JWST was intended to replace the Hubble Telescope which has been in operation for 32 years.

The JWST data will bring us closer to understanding the Big Bang and where the universe came from [at least in scientific terms, not in divine terms]. Taking photos of the deepest reaches of space is like taking snapshots of the past. The JWST is equipped to take pictures in the infrared range giving us clearer pictures of what lies beyond the dusty clouds of outer space.

Since Sputnik was launched by the Russians in 1957, mankind has been littering space with mechanical devices, i.e., space junk, if you will, which is evidence that we humans exist. If our sun should explode, then there is mechanical evidence out there which might tell aliens that we were here, that we have made our mark in the universe.

*****

Discovering Exoplanets

Sam Gregson Interviews
Professor David Kipping

The first exoplanet discovery was in 1991. Now, we have 5,000 accepted exoplanets. We are incredibly biased in our search. We look for stars that are similar to our sun, which is only 10% of what's out there. We are also obsessed with looking for Earth-like planets because that is what we live on. We must get out of this mindset that we are the pinnacle of life, and everything must look exactly like us. [Professor David Kipping]

How do we find exoplanets? The most popular method is "the transit method" where the planet crosses the light of a star and the photometry dips low there. Another method is "the wobble method", actually the first method attempted, where a planet pulls on its star and creates a wobble effect.

"Microlensing" is a technique used for seeing planets that are far away from the Earth. Using this method allows a person to discover stars, planets and other celestial bodies that are thousands of light-years away.

Professor Jason Wright, a foremost researcher for extraterrestrials, has suggested that the natural movement of the stars within the galaxy would be more feasible for getting closer to other stars by just being patient and riding along with the system to get closer to other stars. The star system itself would be the spaceship since we are talking about lightyear distances. We just wait until a system passes by.

"The Aurora Effect", like our lovely "aurora borealis" is a sign for a friendly planet. It's something in the atmosphere that we are used to, with lovely dancing colors, a good sign that there's an emission of oxygen. It might be said that the planet is friendly and probably stable and settled.

When we look for alien life, we look for biosignatures, which we can guess at, if the light shining through the atmosphere shows that there is indeed oxygen, carbon dioxide and methane there. It's amazing what spectrography can detect. This enables us to make an intelligent guess that life exists on an exoplanet.

We also look for "techno-signatures", if there are lights of cities out there or the presence of carbon monoxide indicating a civilization using fossil fuels to drive their vehicles.

Scientific thought of this: that we are the aliens, that an intelligent civilization already colonized Earth in the past without us knowing? One assumes that aliens are greedy, as evidenced in our own human history, to conquer, acquire and assimilate, to take what someone else has. But maybe aliens don't have the same values as we do. Maybe, they are not greedy. Who needs to conquer if they are so advanced that they look on Earth as not even worthwhile plundering?

Science has come a long way in the past 100 years. Not only detecting exoplanets, extrapolating if there is life on those planets, but also what the aliens might look like. Fergus Simpson predicted how big aliens are depending upon the size of their planet and its gravity compared to Earth.

*****

We are an intelligent species, although we must not think too highly of ourselves because we are still listed at Type 0 on the Kardashev Scale, the bottom of the totem pole, with another half dozen or so advanced civilizations above us.

David Kipping and Alex Teachey have surmised that aliens might be able to use a "cloaking device" to hide that they are there by manipulating the dip in a transit of an exoplanet to make us see a straight line in the light reading, instead of the real dip of a transit.

In their article, "A Cloaking Device for Transiting Planets", they said, "We suggest that advanced civilizations could cloak their presence, or deliberately broadcast it, through controlled laser emission." If aliens really wanted to broadcast their presence, they could manipulate the reading on the transit by making a triangle or a hashed sequence, which might be their way of saying hello, similar to the flashing of morse code from the light of a ship to another ship.

*****

Professor David Kipping mentioned "the water hole" frequency in radio transmissions. In SETI programs, the "water hole" refers to the range of signal frequencies between emissions caused by hydrogen and hydroxyl molecules. The reason that most SETI programs choose to listen at microwave radio frequencies is that microwaves are the most energy efficient way to send information.

The waterhole, or water hole, is an especially quiet band of the electromagnetic spectrum between 1420 and 1662 megahertz, corresponding to wavelengths of 21 and 18 centimeters, respectively.

David Kipping hosts the YouTube program, Cool Worlds Lab.

12
New Discoveries

On February 22, 2017, the team of the Spitzer Space Telescope discovered a new exoplanet: TRAPPIST-1e which they say has a liquid ocean. It is 40 light-years or 12 parsecs away from Earth. The star itself is named, TRAPPIST-1 and is located within our Milky Way. The star is a red dwarf, the most common type in our galaxy. It is also within the constellation of Aquarius.

The Trappist-1 star is slightly larger than Jupiter and only 10% the mass of our Sun. Its temperature is much cooler than our sun.

However, because the 7 TRAPPIST planets orbit so closely to their sun, they receive comparable light and heat to Earth. By the way, the life expectancy of a red dwarf, like TRAPPIST-1, is hundreds, maybe thousands of times, longer than our own sun. TRAPPIST-1 will shine ten trillion years, about 700 times longer than the current age of our Universe, while our sun will run out of hydrogen and die in about 5 billion years. Nothing for us to worry about though!

Even though TRAPPIST-1 is located within the Milky Way, it is still too far away to be accessible to us humans. The two planets around Proxima Centauri, which are also within the Milky Way and are 10x closer, would take only 73,000 years for Voyager, for example, to get there!

We need to start planning early, however, if we are to get on a lifeboat to get out of our own Solar System, once our sun dies! Plenty of time, one would think, but not necessarily, when one thinks about the inventions for interstellar space travel that we need to come up with.

Let's hope we have enough time to do this so we can escape our little real estate in space once our sun starts dying.

The main reason, however, that the TRAPPIST system is of such huge interest to us is that its sun has a total of <u>Seven exoplanets</u> orbiting around it, one of them with a liquid ocean. This in, 2017, was a bonanza discovery!

The TRAPPIST-1 star is a red dwarf that is considered "ultracool" and the temperatures of the 1e planet itself is also cool but within range of human endurance, at -17 F to +6 F. The gravity of TRAPPIST-1e is 0.930 surface gravity. Astronomers determine that the planet 1e also has an atmosphere a lot like Earth with clouds!

Astronomers discovered this treasure trove through a method called "transit" where a planet crosses the path of the light from its sun, thus dimming that light enough to indicate that something, probably a planet, was blocking that light. Through spectrographs, astronomers can also determine the makeup of the planet's atmosphere, whether it has oxygen and even water vapor.

*****

Both Jon and Kate were inspired to stay in their field of study when the Spitzer Space team announced the Trappist discovery in 2017. Jon was a reporter for a science magazine, and Kate was an astrophysicist. They had been looking into teaching prior to that but, when new strides were made in exoplanet discovery, they decided to plunge into their work with eager eyes focused on the stars, so they could be part of these new steps into space exploration.

Michael Gillon headed the team of these exoplanet discoveries through Belgium's University of Liege.

The sun itself was named, TRAPPIST-1, and then the planets were named in an alphabetical sequence, TRAPPIST-1b to TRAPPIST-1h. It took several weeks for the team to determine that there were 4 more additional planets to the original 3, that they discovered. These were extra gifts to the astronomers at the Belgium University.

Because these exoplanets have to huddle so closely to their red dwarf star to keep warm, they have very quick years, for example TRAPPIST-1b zips around its sun in 1.5 days and TRAPPIST-1h in 3 weeks. Some year!

Scientists say that "they are likely rocky and watery worlds like our own." They estimate that three of these planets occupy their star's "habitable zone" with liquid water. What a rich find!

13
TRAPPIST-1e

TYPE II, a stellar civilization:

The Trappist civilization had accustomed itself a long time ago to a year's rotation as being only 6 of our days. Maybe, that gave them a work ethic where they got so much more done to control their atmosphere and even the energy from their sun. The TRAPPIST system is 7.6 billion years old, about 3 billion years older than our Solar System. Therefore, their life would have had a much longer time to evolve, creating a very advanced civilization.

The sky in TRAPPIST-**1e** looks pink. The other planets look huge in the sky like red silhouettes, larger than our full moon, and sometimes twice that size, red orbs which cross the pink sky of TRAPPIST-**1e**.

The inhabitants of the planet are used to their planet facing one side to their sun, meaning that the planet is "tidally locked" to their sun, like our moon is to us. Both the dwellings and the physiology of Trappists had developed to accommodate the quirks of their one-sided planet, a planet that is not only one sided, but that also orbits their sun in 6 of our Earthly days.

Astronomers suggest that a one-sided planet does not mean there is no life there. A "terminator zone" is a band between the hot and cold halves, where the climate might just be temperate enough to sustain life.

The other planet which falls in line with 1e, is 1d. It too has developed an advanced civilization, almost as advanced as **1e**. Both races are very similar in looks and size.

However, **1d** falls just within the inner edge of the habitable zone. It is the tiniest of the series, around three-quarters the size of the Earth with a crowded population. One year lasts 4 days. **1d** is jealous of **1e** because **1e** is larger, almost the size of Earth, and therefore, has more real-estate. **1e** is also situated more comfortably within that habitable zone. One year there lasts 6 days.

People are dark and thick skinned on both planets, with a fast and strong heartbeat to keep them warm in their cool worlds. They are a hearty breed and live to a thousand years, i.e., of our years! They are planets of Methuselahs!

Trappists, on both 1d and 1e, are small and stocky in body. Their cranial cavity is twice that of the human skull. We would be like monkeys compared to their intellectual grasp of physics, astronomy, and relativity. To harness energy from their star, TRAPPIST-1, these beings have built a Dyson Sphere around their sun from which they could extract photons to power their various instruments, devices and starships. Their mode of interstellar travel is "Solar-Sails" which are flat constructs that capture the photons from stars to propel them through space, like an advanced version of the old sailing ships which used wind to traverse the high seas on Earth.

The Trappists mastered the weather on their planet eons ago, constructing huge warming centers in the four corners of their world to keep their ambient temperature warm at 75 F. They made a few exceptions for recreation, where snow making machines have created great slopes for skiing.

Earth would take another million years to develop into a civilization equal to that of the Trappists, which on the Kardashev Scale was Type II, a stellar civilization.

Since a war almost broke out between the Trappists on 1d and those on the larger planet 1e, a civilized agreement was reached some thousands of years ago where 1e gave 1d the technology and manpower to make the planet on the outside of their habitable zone, i.e., Trappist-1f, more habitable, so that the overflow of people from 1d could expand to it. All three planets agreed to some form of birth control in commensurate with the real-estate of their planets.

Lor, a respected astronomer and specialist in ethics, was uneasy about this concession among all the Trappists of every stripe from 1d to 1f. There were factions on 1e who considered themselves purists, who felt uncomfortable extracting populations from 1d and exporting them to 1f. The purists on 1e felt sandwiched, surrounded, as it were, figuring that the other two planets should simply have become extinct. But the situation was as it was, even among these super intelligent beings. They mastered the weather on their planets, but they still had physical bodies and thought in terms of dwellings and real-estate.

Lor wondered if a precedent had been set within the TRAPPIST system for a cataclysmic war in the future. Lor's only solution was birth control and living within one's means.

The Trappists were not the only ones who faced such dilemmas. Other advanced civilizations had already arisen in other systems of distant stars, even on those planets outside of the familiar Milky Way Galaxy. They needed to come up with policies for managing similar problems, like the overpopulation of TRAPPIST-**1d**.

But more importantly, they needed to adopt firm policies in dealing with alien species that were lower in technology than they were. The Trappists agreed, finally, that they, along with species in other galaxies, would make it an interplanetary law never to interfere with primitive societies.

Earth was on their radar, so to speak, but it was hardly worthwhile that any of the TRAPPIST civilizations noticed that backward planet. Earth was where TRAPPISTS were some million years ago. So, Earthlings had a long way to go to make first contact. What kind of meaningful conversation could one have with monkeys?

Lor told his fellow astronomers that Earthlings were too egotistically immature to harness technology such as their own, something that was too far above them. If their own Trappist system had just barely forged a peace within their own species, how much less could hot-tempered Earthlings agree to any interstellar peace?

Lor had been monitoring "the busy little bees" on Earth for several centuries. He was amused. However, he looked at them with more respect, when they launched the James Webb Telescope, to see what was beyond their own galaxy.

"They are a curious race. Should we say hello to them?" asked Lor of his fellow Trappists. They smiled and said No, not yet. Not yet, could mean six days, six years or it could mean a million years from now.

*****

All in all, it would not matter anyway. The Trappists certainly could not give Earth any technology that it was not ready to receive and maintain.

Directive Number One must be followed. It was an agreement among all advanced civilizations, all across several galaxies. *Not to interfere with the development of primitive species!*

Lor was convinced that Earthlings were so instinctively bestial that giving them the secret to nuclear fusion would mean they'd burn or kill themselves. They had split the atom already in nuclear fission and then nearly destroyed their world in a nuclear holocaust. No, it was better to leave them alone. Even to hide from them.

"We learned in our own history," said Lor, "that there is no cure for stupidity."

14
The Watering Hole

Type 0 advanced, our Earth:

The National Aeronautics and Space Administration, aka NASA, is located in Washington D.C. The leadership works there at the low-rise building in the two-building Independence Square complex off 300 E Street SW.

A special conference was held there that week. The 5,000th discovery of an exoplanet marked a special milestone for SETI, the Search for Extraterrestrial Intelligence. The discoverers of that planet, plus a number of keynote speakers, famous for making new discoveries in outer space, were invited to the celebration.

Kate Wilson had participated to some degree in the planet's discovery. Even though she hated celebrity status and celebrations of any sort, she accepted the invitation anyway.

At the same time, Seth, a pub owner in the city, celebrated his first-year anniversary as a successful entrepreneur. He had seen the need for a pub, an informal relaxing place for employees at NASA to come after work. Within the year, it had become a popular drop-in center for NASA employees, astronomers, astronauts and even their leadership.

Seth thought about naming the place, "The Water Hole", but didn't think it sounded quite right "The Watering Hole" sounded much better.

He got the idea from a YouTube episode where Sam Gregson interviewed astronomer and Professor, David Kipping. Dr. Kipping explained that SETI was using a radio frequency that was particularly quiet, a radio frequency that was most likely used by aliens to broadcast that other life existed in the galaxy.

SETI called their program, "The Water Hole", taking the idea from the African steppes where a variety of animals shared a common watering hole to survive.

The owner liked the idea, a good name for a pub. "The Watering Hole" where the variety of employees at NASA could quench their thirst. The name was adopted and became a preferred place for NASA employees to sit, drink, eat and relax after work. It was a good way to wind down from the serious briefings and conferences held by NASA.

*****

Kate Wilson walked into "The Watering Hole". She was hungry and thirsty. She had put in overtime in a debriefing session. Her specialty was reading the spectrographs of exoplanets. She had acquired her B.Sc. at Columbia University, and her Master's degree and Ph.D. from Johns Hopkins. She was a regular that week at "The Watering Hole".

Seth, the owner, asked her, "What's new?"

Kate replied, "NASA's excited about the 5,000th exoplanet we discovered out there in the Goldilocks Zone. They sent me the spectrographs and want me to analyze the data...by yesterday."

"Everybody's in a hurry?" said Seth. He added a placating word to his customer. "I suppose someday, you'll be the first to discover new life out there."

"It's only a matter of time," said Kate.

Seth brought over Kate's order of Fish 'n Chips. She sat there dipping her French fries into the ketchup staring into empty space after a tiring day at work at the conference.

Jon Hart, a news reporter, joined her and ordered the same, Fish 'n Chips. In the meantime, he took one of her fries and dipped it into her ketchup. She nodded assent and said, in jest, "Permission granted."

"Nice that NASA pays all the bills for this discovery," said Jon.

He wasn't your run of the mill reporter. He was a "science reporter" and for that time, Kate's significant other. He had known her for 5 years. Friends originally thought they'd marry, but as the years flew by, they had given up on that idea...and so had Kate.

Jon worked for a magazine called <u>Astronomy Now!</u> It was a glossy magazine which featured UpToDate articles on the latest discoveries in outer space.

Being Kate's boyfriend meant he had the inside scoop on the latest outer space story, which in this case was the discovery of the 5,000th exoplanet. Kate was always good for a golden quote, "One of these days we've got to hit the jackpot and discover life on one of those planets."

The advantage to being a freelance reporter included a flexible schedule where Jon could come along to Kate's conferences where he was assured of getting exciting stories for his magazine. The editor gave him huge leeway, being keen on how networking worked.

Jon and Kate made a strange couple. Usually, one would assume that like would attract like. However, astronomer and reporter, they somehow clicked. They got along and truly loved each other.

They'd been engaged for about a year, the same time that Seth had opened his pub. But the engagement ring is as far as this relationship apparently went. Jon could not agree on a date to show up at the altar. He said to her, "Don't get me wrong, Kate, but give me some time. I'll let you know when I'm ready." Readiness never seemed to come.

Kate's co-workers still called her Kathleen, but ever since Jon regularly stayed over at her place, he called her by the more familiar, Kate. It sounded more romantic.

There were only three customers in "The Watering Hole" at supper time. Jon and Kate sat at one table enjoying their Fish 'n Chips. A big fellow, huddled in the last table, was trying to sink into his seat with a book.

When Jon got up to go to the washroom at the back, he deliberately eyed the fellow's reading material.

Jon excused himself. "Sorry but I'm always interested in what other people are reading," he said. The big fellow looked up with a friendly smile. He held up the book with the title clearly visible, <u>The Elegant Universe</u> by Brian Greene. The big fellow commented, "I try to read what's appropriate in whatever venue I'm in. In this case, I understand that there's a conference in town to celebrate the 5,000th discovery of an exoplanet."

"My fiancée is one of the astronomers invited to that celebration," said Jon.

The big fellow introduced himself. "My name is Jack. I'm retired from the army. A former Major with the Military Police. Right now, I'm enjoying my retirement, seeing the sights of our grand country."

Jon appraised the big fellow in a new light. "I've read that book too. I'm glad that physicists like Brian Greene have made our universe more understandable to people like me. I write for a science magazine."

When Jon came back from the washroom, he invited the big fellow to come over and sit with him and Kate since there were only the three of them in The Watering Hole. The regular crowd had not started to come in yet before 6:00 p.m.

Kate commented that she was impressed by a civilian reading <u>The Elegant Universe</u>.

Jack replied, "You have to have an inquisitive mind. I'm always intrigued by the possibility of life in other galaxies. If I could travel at the speed of light, I'd be a hitchhiker in the universe, instead of thumbing my way across the United States. It's a wonderful time to be alive."

Jack ordered Fish 'n Chips since that seemed to be the popular fare in the establishment.

15
String Theory

Physicist, Brian Greene, suggests in a YouTube interview that he'd feel more comfortable with The String Theory actually being called "The String Hypothesis". He doesn't feel that the concept of energy strings hasn't quite reached the level of a theory yet. He has strong reinforcements for his idea in other physicists like Michio Kaku. The two are formidable exponents of the idea, clear and articulate speakers when they explain it.

YouTube: "Why is our universe fine-tuned for life? Brian Greene"

String theory says that if you go smaller and smaller "you'd find something else inside these particles [like electrons and quarks], a little tiny filament of energy, a little tiny vibrating string! Just like a violin, they can vibrate in different patterns, producing different musical notes. These vibrating strings, when they vibrate in different patterns produce different particles. So, electrons, quarks, neutrinos, photons would be united into a single framework, as they would all arise from vibrating strings…It's a cosmic symphony where all the richness that we see in the world around us emerges from the music that these little, tiny strings can play."

*****

Brian Greene, however, admits that the theory has inconsistencies, unless we allow for something unusual, extra dimensions of space!

These extra dimensions are folded up and so tiny that we cannot see them.

> The shape of the extra dimensions constrains how the strings can vibrate. And in string theory, vibration determines everything. So, particle masses, the strengths of forces, and most importantly, the amount of dark energy would be determined by the shape of the extra dimensions.

The challenge is that we don't know what shape these extra dimensions can take. They are so tiny.

Nevertheless, both Brian Greene and Michio Kaku say that this theory is, "the only game in town", because mathematically, it works.

In string theory, spacetime is 10 dimensional, 9 spatial and one of time. In 1995, Edward Witten conjectured the existence of M-theory, which gave physics 11 dimensions and the plausibility of a multiverse.

*****

YouTube: "Where are all the Hidden Dimensions?"

Theodor Kaluza, a Silesian [Prussian] physicist asked himself, what if Einstein's equations would be written out for 4 spacetime and 1 time dimension, instead of the 3 conventional dimensions? This was in 1921!

Kaluza argued that the extra dimension was so small that it was not observable. Kaluza solved the equations indicating gravity and the presence of an electromagnetic force. His paper was sponsored by Albert Einstein himself.

However, World War II and a general disinterest in Kaluza's ideas made the idea of extra dimensions ignored.

The resurgence of string theory did not occur until the mid-1980s when Michael Green and John Schwarz at a conference at the Center for Physics in Aspen, Colorado, suggested 6 extra dimensions for solving inconsistencies in string theory. They suggested that these extra dimensions existed in tiny, tiny dimensions of spacetime called, calabi-yau manifolds. That discovery in 1984 revivified string theory for two, now famous physicists, Brian Greene who was then 21 years old and Michio Kaku who was 37. The calabi-yau geometries satisfy the equations of Einstein's theory of gravity.

There is a problem here, that Einstein's equations are classical; they do not include any quantum physics. Classical equations fall apart the moment quantum effects are introduced. The quantum leads to uncontrolled infinite amounts of extra energy that wrecks the simple classical solution. The protection against such inconsistency is adding an extra symmetry called, "supersymmetry", which guards against quantum effects. Calabi-yau manifolds are the solution in supersymmetry.

*****

When Jack got up to leave, after his meal, both Jon and Kate were impressed with the man's size. At least, 6 foot 5 inches tall, a good 250 pounds with a 50-inch chest. A beefy athlete who had a mind for "the elegant universe".

Jack enjoyed not fitting into the stereotype of the big, dumb brute. He confessed, "Sometimes, I play dumb just for the fun of it. I understand most of the concepts in the book about string theory and the 11 extra dimensions."

Kate was impressed by their new friend who didn't fall into the typical stereotype of a big, dumb guy.

"Bright fellow," she said, "behind the Neanderthal exterior."

Jack found paradoxes amusing. It was ironic that he was reading a book entitled, The Elegant Universe, whereas his life, most often, involved violence and fist fighting. He didn't look for it; somehow, trouble always found him. He told Kate, "We are a most inelegant species within this elegant universe. Maybe Brian Greene should have noted that irony in his introduction."

16
The Next Day

Like Jack's antithetical makeup, Jon too had interesting opposites within him. He put scientific concepts into layman's language and did a good job of it. He was very adept in his clear writing style and respected among journalists and scientists alike. But like Jack, Jon was a paradox. He liked watching old Audie Murphy Westerns to relax from the cerebral demands of his job.

He was glued to the TV screen one evening watching, "The Posse from Hell", over a Swanson TV dinner. That was his hurried life as a bachelor when he did not have time to make his back-up meal of tomato soup and grilled cheese sandwiches.

That's when Kate yanked him out of his apartment to grab a more substantial meal at "The Watering Hole".

They went the next day too. Seth's pub featured lasagna that evening. That and the beer sounded good.

When they entered the pub, they saw Jack sitting in his usual corner, quietly reading <u>The Elegant Universe</u>. He was almost done.

Jon invited him over. He smiled, closed his book and sidled over to the table where Kate welcomed their new friend.

She joked, "I bet if there was a Restaurant at the End of the Universe, you'd be there, but only if they had good strong coffee." Jack never ordered a beer with his meal. He preferred his coffee.

Jack, smiled and conceded, "You read me like an open book." Kate had already ordered a carafe of coffee for the table.

"So, you're nearly done that book?" she asked pointing to <u>The Elegant Universe</u>.

"Brian Greene has a knack for explaining the unexplainable," commented Jack., adding, "It's been an enjoyable read."

The trio concentrated on their lasagna meals. Jon and Kate sipped their beer; Jack sipped his steaming coffee. They were quiet enjoying their food.

The only disturbance came with the rumbling sound of motorcycles outside the premises. Three broadly built bikers opened the door. They looked around and nodded to each other. Several of the other patrons didn't like the looks of this and sank further into their seats, staring at their plates heads down.

One of the bikers said, loud enough so everyone could hear, "Is this where we find the geeks who are wasting our tax dollars?"

Jon and Kate tried to ignore them. Jon was of average height at 5 foot 8 inches tall, compared to the monsters who just walked into "The Watering Hole", each one over 6 foot holding up a beefy 200 pounds.

Jon had studied books, not boxing. He was intimidated. Then he was frightened when one of the bikers grabbed him by the scruff of his neck and yanked him erect.

Jack stood up. "You want to try that on me?"

The biker dropped Jon back down.

"The three of us should be able to take this overgrown ape," said the smallest of the bikers. He always relied on numbers to back him up.

"Let's be civil about this," said Jack. "How about if we take this matter outside. I can see now why Earthlings are relegated to Type 0 on the Kardashev Scale."

"What the hell are you talking about?" challenged another of the bikers.

"Just that if we are going to act like apes, we'd better take this outside."

The four of them exited the premises. Three bikers surrounded Jack who now stood in the centroid of an equilateral triangle. He was relaxed. The trio was wound up tight and ready to spring.

Seth, Kate and Jon stood open-mouthed by the window. The other customers sat at their tables still staring at their food. They did not have the courage to stand up and leave.

Somebody asked, "Were the police called?" Seth had already done that, expecting the worst to break out outside his pub with three guys against one. He would have welcomed the bikers, if they had actually come in for food and acted in a civil manner.

What Seth, Kate and Jon saw astounded them. The tallest of the trio had armed himself with brass knuckles. He was also the mouthpiece for the group. Jack assessed him as the alpha male. Take out the leader and the pack would cower.

Jack let him step in. Big mistake on the biker's part. Jack sent out a boot, a quick roundhouse, into the man's face. The biker reeled back stunned. By this time, Jack came closer to the next tallest biker, giving him a quick jab with the heel of his hand. Jack heard a jaw crunch. The fellow dropped shaking his head to clear the stars. The smallest fellow turned and ran. No more big talk.

Moments later, a cruiser showed up with the biker who ran, sitting disconsolately in the back seat. Another cruiser made room for the other two bikers who didn't look so happy either. Their motorcycles were impounded.

Seth told the police, "They got what they deserved. They threatened my customers."

Jack had intended to hitch-hike out of town the next morning, a usual routine, but he stayed to make sure the police got his statement.

"I'll sign the complaint," he said. "If those three clowns say any different, then there's enough customers and my friends here to back me up. They were asking for trouble."

The police agreed with Jack. "Eyewitnesses back you up. These guys have no vehicle insurance. They are wanted in another state for assault."

When Jack came back into the pub, the patrons clapped their hands in approval for what Jack had done and congratulated him on standing up against the bullies.

Jon knew all about that. He had been beaten up enough times in grade-school because he was a geek and smarter than his peers. He wore glasses and even as an adult looked like the epitome of a geek. Kate loved him anyway.

17
The Day After

Jon and Kate sat in "The Watering Hole" alone. No Jack. He had disappeared, probably hitchhiked out of town. Probably on the way to another town where he would right the wrongs there, take care of bullies for which he had a knack.

"Interesting guy," said Kate.

"Am I supposed to get jealous?" asked Jon.

Kate laughed. "There is no reason to be. You are my geek hero and my number one." She patted his hand. She thought, however, that he would show more courage by actually setting a date for their wedding.

Jon had always harboured phobias since he was a kid. Maybe, that was what came from being a geek and somewhat autistic. He had been tested and been told, "Yes, you are on the spectrum, but don't worry about it."

The phobia which plagued Jon the most was the disaster that always lurked around every corner. He feared things. He feared the worst. Maybe, he got that from his mother, who being European, had a knack for being pessimistic and seeing black in everything.

Maybe my fears were real, he thought. A car could come crashing into him at a green light anytime. Lightning could strike. Kate could fall in love with a hulk like Jack and leave him.

But none of the above were really true. Kate told him, "Where will I find another you? You're a keeper; you belong to me." Jon liked that idea, but he froze when he thought about marriage and real commitment.

Their love of astronomy was not enough to keep them together. The stars were not a sufficient bond, with Kate finding the stars and Jon writing about them.

Sure, there were other things. They both loved classical music and played instruments. He the violin, and she, diminutive as she was, the viola.

Jon used to joke with her. What was the difference between a viola and a violin? A viola takes longer to burn. Kate had punched him in the shoulder when he told the joke, but not too hard.

They both joined a local community orchestra and enjoyed the camaraderie. "Beats stamp collecting," joked Jon.

Often, they played in seniors' homes. They felt that their Christmas and Easter concerts were a very positive thing to add to a society. It made them better people. Besides, they were treated to coffee and cake after the concert.

*****

Kate had been assigned to work on the team which launched the James Webb Telescope, one million miles away from Earth. The big event was scheduled for Christmas Day, December 25, 2021.

Jon was told by his editor, "You have a free reign. Bring me articles on this new telescope which will keep our readership glued to their seats."

There were 18 individual mirrors which made up the primary mirror for the James Webb Telescope. Its greatly improved infrared resolution would allow it to view objects too faint for the old Hubble Telescope to see. The most distant galaxy ever observed was GN-Z11 located about 32 <u>billion</u> light-years away.

At about 46 billion light-years away from Earth, we reach the surface of "last scattering". What lies beyond the observable universe is the "opaque universe". And what lies beyond that, who knows?

According to Einstein's reckoning, a flat universe must be infinite. That's ironic because people in the Middle Ages conjectured that the Earth was flat...and of course, it did not turn out to be so. But a flat universe is different and real.

Both Jon and Kate had been drawn into astronomy with just such questions. What was...or rather is out there?

*****

By July of 2022, the telescope was scheduled to send its first pictures of intergalactic space back to Earth. Jon and Kate were devastated when a meteorite smashed into one of the mirrors in May, thus creating a blind spot in that portion. But scientists were able to compensate digitally for the accident, thus regaining sight of distant stars and galaxies which still was better than Hubble's weaker eyesight.

Space was a dangerous place. Especially, when objects swirl and dance and sometimes collide with each other.

"I can't see that sort of collision happening when Andromeda collides and merges with our Milky Way," said Kate.

Kate added, "Andromeda should win that one since it's twice the size of our Milky Way. But distances are so far away in their constituent components, that collisions should be rare indeed even between Andromeda and the Milky Way."

As an afterthought, she also added, "Anyway, we won't be around in 5 billion years to see it."

18
Jon and Kate

Jon and Kate had met on a canoeing trip in Algonquin Park five summers ago. They both answered the same ad, and felt that maybe the stars had something to do with it.

At first Kate was not impressed with Jon. She was two inches taller at 5' 10". He barely squeezed into the average range, at 5' 8". He was slight of build and wore spectacles. A sort of replica of Woody Allen. He had a knack with words though and often said funny things which Kate found charming.

Kate took to canoeing with no problem. Along with her mind for math, she was also athletic. Somehow, things came easily to her. She had always balanced her job in astronomy with her athleticism. Kate had been a basketball star in high school and was head of the chess club. She was a "scholar-athlete".

Jon was far from that. His gift was writing, having written a secret stash of poetry and the first chapter or two of a novel. He liked outer space and felt that there were an infinite number of themes there among the stars.

When Jon was stymied in a writer's block, his editor told him to get into the great outdoors for a week or two, to clear his head. His editor at "Astronomy Now!", valued Jon's work, not only for the catchy way he put phrases together but also for his sheer output. About 30% of the Magazine's articles were from Jon's hand. He was also the Magazine's photographer, supplying his editor with superb pictures of up-and-coming astronomers peering through telescopes. Jon did not take your run of the mill head and shoulder photos.

"You're doing a great job," said the editor, "take some time off. Go canoeing."

"I don't know how to canoe," rejoindered Jon.

"Learn," said the editor.

A week later, Jon stood at a wharf in Algonquin Park with a dozen other awkward looking people wondering why they ever signed up for something where they had to puff and pant all day long. Jon was not a strong paddler, nor a strong swimmer. He tightened his life jacket.

By chance, the leader of the group paired Jon off with Kate. Kate looked like she knew what she was doing. She said, "I'll steer." Jon said, "Fine by me."

They didn't talk the first day. Jon was glad when the leader veered into a little cove by late afternoon. They were ready to set up camp. Jon's shoulders ached, as did his seat.

The idea of the excursion was to fish for their supper on the first day. The group did well and soon several Bass were frying in two pans. The leader had brought pop and Gatorade to drink. Jon and Kate enjoyed the meal cooked over a Coleman stove.

After supper, the leader asked each one of them to give a run-down of who they were and why they had signed up for this grueling excursion in Algonquin Park.

Kate confessed she was a student at Harvard, a candidate for her Ph.D. "I'm Kate, an astronomer, and stars have always interested me," she said. "My specialty is searching for exoplanets." She said that canoeing was a good way to relax.

Jon thought, "Relax?" The mere thought of paddling tired him out.

He said, "My name is Jon. I'm a writer. I work for a magazine called, <u>Astronomy Now</u>!" He confessed to writer's block and hoped the trip would cure that problem.

When the group roasted marshmallows, Kate disappeared down to the water. The sun had already set. She found a log and perched there looking up at the stars. Jon came over to join her.

"Do you mind?" he asked.

She moved over.

"I feel closer to the stars, than I do to people," she said.

"I often have nothing in common with people, myself," said Jon.

Before the weeklong trip was over, Kate decided she liked Jon enough to have a friendship with him. They lived close enough together in Cambridge, Massachusetts to see each other on a regular basis. They went on other canoe trips together.

Soon they were an item, boyfriend and girlfriend, but that's where their relationship seemed to get stuck. Kate felt that being boyfriend and girlfriend sounded too adolescent.

She was a woman with a career, no longer a student, nor a kid. Being a scientist, she felt that the natural order of things was "evolution" of some sort. A date for a wedding never materialized.

*****

Jon seemed comfortable with their "arrangement" where he often spent weekends over at her place, where he had a reliable source for his articles, and where he had a reliable canoeing partner.

Jon had a problem with commitment, and he shied away from the topic of children. He liked his freedom, and he liked being friends with benefits. Nothing to tie him down. And so, what Kate wanted secretly to say to Jon just faded away, and the situation just went on with that silent "arrangement".

19
They Are Already Here!

There was a strange interstellar object which crossed into the inner solar system in October 2017, discovered by Robert Weryk using the Pan-STARRS telescope in Hawaii. When it was first observed, it was 21 million miles from Earth. The object was estimated to be cigar shaped, about 3,000 feet long and 548 feet thick, red in color. It was named "Oumuamua", which in Hawaiian means "scout".

Although many astronomers label this phenomenon as one of the natural things floating around in outer space, a few had conjectured that Oumuamua could just be space junk from another civilization.

Author, Avi Loeb, suggested that this long shape could be a "LightSail" driven by sunlight. It might be a space buoy sent out by an ancient civilization which itself was looking for other life.

Avi Loeb said that we missed a good chance to find out for sure by not looking closely enough at the object while we had a chance, as it flitted through our solar system, and then into interstellar space, out of our reach. He based his claim on the object not having a tail like a regular comet and also that it was able to change direction unexpectedly in its trajectory. Why would it do that unless it was steered in some way?

Professor Loeb cited Winston Churchill who gave his opinion in an essay written in 1939, entitled, <u>Are We Alone in Space?</u> The essay sat unpublished for years, his thoughts ignored: "I am not sufficiently conceited to think that my sun is the only one with a family of planets." The great statesman went on to say, "I [do not think that] we are the only spot in this immense universe which contains living, thinking creatures, or that we are the highest type of mental and physical development which has ever appeared in the vast compass of space and time."

Well, we missed our chance with further investigation into this interloper which transgressed into our solar system in 2017. Yet, the incident raised some valid speculations about what is out there.

It is estimated that the cost of WW II was a whopping 1.3 trillion dollars which is 18 trillion dollars in today's currency.

Professor Loeb asks, "What if humanity of the 1940s spent that money, instead on the exploration of space?" Professor Loeb further asks, "What investments are we making to ensure our continued survival after the inevitable death of our Sun in 7 billion years?" Of course, human nature tends to put things off, but it's a good question.

In all of time and space, there might be civilizations which wink in and out of existence. We might not be impervious to go the way of the dinosaur. Professor Loeb wrote his book in 2021, <u>extraterrestrial, the first sign of life beyond Earth</u>, a worthwhile read.

Professor Loeb suggests that we are likely to be at the center of the bell curve in universal intelligence, rather than at the higher end of the intelligence curve, where we might smugly place ourselves.

For our civilization to mature we need to venture into space and seek others. If we discovered extraterrestrial life, new fields of study would open up, such as Astro-linguistics, Astro-law, interplanetary law, Astro-ethics and Astro-technology. Finding extraterrestrial life would impact not only science but also education and religion. Lots of new things to grasp!

We are not new to thoughts about life in outer space. Author, Wade Roush, traces the history in his book, Extraterrestrials, two and a half thousand years to Ancient Greece. Philosopher, Anaxagoras, was banished for his thoughts that the moon was a great rock and the sun a hot rock. Anaxagoras entertained the idea that the moon might be inhabited. Democritus posited that there were an infinite number of atoms, and therefore, said Anaxagoras, there were an infinite number of worlds.

Jump to the Middle Ages, the Dominican friar, Giordano Bruno, published his speculations between 1584 to 1591, claiming that stars were distant suns and that these suns had planets which might be inhabited like Earth. It was his interest in magic which gave the Church an excuse to burn him at the stake in 1600.

Jump to 1947, at Roswell, New Mexico, where a recovery of a balloon sparked the Unidentified Flying Object, "the UFO craze". Yet, the crash of a balloon did not explain the sightings of saucer-like flying objects which could zig-zag and go 20 times the speed of sound.

Astrophysicist, Michio Kaku, says that if these are from extraterrestrials, that they probably are not interested in us, like we might not be interested in holding a conversation with squirrels. We would lose interest. "The squirrel has nothing to offer you." Professor Kaku leaves open the possibility however, that extraterrestrials are already among us and that they have been studying us for decades. He suggests that humanity should not announce our presence to outer space, but that we need to find out what their intentions are first.

Professor Kaku says that we need to revamp the outer space treaty of 1967. It is old and out of date. The original idea was, *#1. No nuclear weapons in outer space and #2. Celestial bodies cannot be claimed by any government.*

Modern weapons were not foreseen by the 1967 treaty, such as laser beams. Also, you don't have to be a country to send a rocket to outer space. Elon Musk, for example, is a private investor involved in exploring and possibly exploiting outer space. Would he stop at planting a personal flag on Mars? "We need a revision of the 1967 treaty," asserts Professor Kaku.

The professor also says that there is a death warrant on the human race from several sides, never mind any threats from aliens.

Right now, the space race among the three superpowers, Russia, China and the United States, poses a huge threat to world peace.

But later on, in 10,000 years, mankind will face another ice age which might send us into oblivion. Then in 10 million years, we have to worry about another asteroid impact, such as the one that destroyed the dinosaur 66 million years ago. Lastly, in 5 billion years, the sun will eat up the earth.

The sun itself will end in ice as its fuel is burnt up; then the universe also will end in ice when it expands to the point where temperatures reach absolute zero. "Physics has a death warrant for all life in the universe," says Dr. Kaku.

*****

Professor Michio Kaku hopes we will eventually have the power of the Unified Field Theory, the power of String Theory, and the power to manipulate gravity, as well as space and time. At that time, we may want to create a "lifeboat" to leave our universe and to travel to another universe where it's warmer, another parallel universe, so that we will have another universe to mess up.

20
Another Day

Jon and Kate missed their little chats with Jack at "The Watering Hole". They wondered if he ever made it to Virginia. Last they heard, he intended to track down a sexy voice there whom he heard over the telephone. Like other people, he was searching for that impossible dream, for that elusive someone who would fill his own personal void. Someone to love.

Jon was happy he had found Kate. They shared a quiet dinner of Fish n' Chips. Jon sipped his beer; Kate savored her white wine. Did white wine go with Fish 'n Chips? They did not adhere to high-brow protocol.

Their attention was drawn to the TV screen as breaking news came in about another incident of an honor-killing in the United States. That sort of news had abated in the last few years. It tied in neatly with what happened to Malala Yousufzai a decade ago.

Nothing much other news had surfaced, neither in the States, nor the Middle East, after Malala had been shot in the head by the Taliban in 2012.

Maybe, things settled down since the Taliban had taken over the government of Afghanistan in the summer of 2021. The news mentioned Malala and the fact that she had earned a degree from Oxford University in philosophy, politics and economics since those early days. She had been outspoken about women's rights in the Muslim world and won a Nobel Peace prize for that effort. However, in Pakistan, said the new report, in Malala's home country, gender equality was rated low, where women were relegated only to domestic help. Most women still wore the hijab or burqa. In Afghanistan, where the Taliban ran the government, the hijab was compulsory.

Everybody knew these things, but no one was expecting another honor-killing out of the blue in the United States. Somebody in the pub commented, "Keep those people out of America with their crazy religion."

The latest incident of an honor-killing showed that extreme ISIS had not mellowed.

But maybe, neither had the red necks in the United States. Mass murders kept happening, some of the most shameful at schools.

The 2018 shooting at the Marjory Stoneman Douglas High School in Parkland, Florida was horrific, costing the lives of 17 people. Nikolas Cruz, a former student himself, was described as cold and calculating. Four years after the event, the sentencing stage of the trial was still going on in the courts! Gun control remained stymied in Congress, with the National Rifle Association and gun advocates standing up for their historical right to bear arms.

Jon, Kate and Seth shook their collective heads as they watched the news from the overhead TV in the pub. But they were more concerned about other things in their lives. Seth about the success of his pub. Jon and Kate were nervous about a new space launch, which was just a communications satellite, nothing grand. They hoped that no incident would occur there, a good place for extremists to make some stupid point.

They had already bought their airplane tickets go get back to work at the Kennedy Air Center in Florida. They enjoyed their little condo there and the chances where they could just bicycle in with binoculars and cameras to witness space launches.

They had powerful binoculars for these space launches which were becoming as popular as bird watching to serious "birders". They also bought comfortable camping chairs which could be folded up neatly to be carried on the back of their bicycles or in the trunk of their car. Often a sandwich and a juice sufficed for these outings.

They had been privileged to watch the James Webb Telescope launch in December of 2021. They anticipated the first great photos to come back to Earth on July 12th, 2022.

How many more galaxies were out there that Hubble hadn't even seen? How big was our universe? And Jon wondered, "How big is the God that made all this?"

Jon was a believer, but he leaned towards his own idea of God, or rather his own impression of a deity, not the anthropomorphic type that most religions depict. At least, Islam did not put an image to an indescribable God. You dare not put a face on the indescribable.

The Catholics had a typical image of Jesus kneeling in the Garden of Gethsemane. He was a handsome Caucasian type, with a straight nose and a well-kept beard and long hair, like a well-groomed hippie. This did not sit right with Jon. "The Father" was depicted by Michelangelo in the Sistine Chapel as a white bearded man with his finger extended to touch Adam in the act of creation. Lovely to look at, but hardly believable.

Despite these stereotypes which appealed to the public, Jon felt he needed to believe in something, but something beyond typical images, especially in times of need or at the scary moment of death.

"Forgive my weakness," he prayed to his silent deity. "Maybe it's a delusion," he acknowledged, "but one that I need as I lay me down to sleep."

*****

Jon and Kate ordered another beer to wind down the evening at the pub. They were still watching the news.

Another item came in. The United States had granted Ukraine another great sum of money to fight the Russians. Originally, Putin claimed that the Russian tanks aligned at the Ukrainian border were there simply for "military exercises". Hitler had claimed the same thing just before his army invaded Czechoslovakia. The world was surprised that Ukraine, under Volodymyr Zelensky, had lasted this long since the Russians first invaded several months ago. Money and weapons sent to them from Western allies kept them going to stave off the Russian Bear.

The whole thing had escalated some months previously when Ukraine made overtures to the West to become a member of NATO. Russian President Putin warned that Ukraine must not do this, thereby interfering with a nation's sovereignty.

It seemed like the war between Ukraine and Russia would never end. Old buildings had been destroyed; women and children killed. Power and politics were reminiscent of the Second World War.

*****

Seth, owner of The Watering Hole, finally said, "I'm closing up." He turned off the TV. He shook his head about the recurring theme of bad news.

He knew that he would not see Jon or Kate again for a long time, maybe until the next Conference. He commented jokingly, "Did you hear the one about aliens landing on Earth looking for intelligent life. I told them to move on." Seth made a closing remark as Jon and Kate walked out the door, "By the way, e-mail me from Florida once you get there. Let me know how you're doing." Jon gave his friend a thumbs up.

21
Unsettling Space Launch

Jon and Kate were settled on the Cocoa Beach Pier close to the launch site in Florida. They had their lawn chairs, their binoculars and their lunch and juice. They were waiting for the count down. It wasn't a hugely important rocket launch. Just another communications satellite.

A shot rang out. Nobody seemed hit. Several Muslim militants demanded that people gather around in a small group. The leader announced a manifesto, "Allah does not want man in space. Man is not meant to go up there to the heavens. That is Allah's domain, forbidden to us." Three of his friends herded the crowd into a tight-knit unit. They stressed that nobody would be hurt, if the launch was called off.

It took only 10 minutes for the center to call off the launch and concede to this demand by the terrorists. As General Hunter said, "Give them what they want. We can always reschedule a launch for another day. No big deal."

When the SWAT team moved in, the terrorists gave up but not without an incident. They had been heard, and at least for now, Allah would be pleased with their efforts. They were happy about that, but they did not count on a "cowboy" in the crowd pulling out his own gun and shooting back at them.

A beefy red-neck with a baseball cap sporting the American flag, pulled out a concealed pistol and fired at one of the gun-toting Jihadists.

"It was self-defense," he later claimed. He shot the man dead, but one of his bullets went stray, and struck Kate in the shoulder. For a moment, the situation looked like something out of the Wild West. Luckily, when the terrorist fired back, one of his bullets hit the trunk of a tree, instead of another bystander.

"It was him or me," claimed the red neck. He looked down at his victim who still had his eyes open, with a third eye staring back at him in the middle of his forehead.

"Nice shooting," said his buddy. The red neck smiled. The news reporter asked if he had ever taken target practice with that clear shot to the forehead.

"Hell no," said the red neck, "who needs it? All you do is aim your gun like it was your finger and then just squeeze the trigger."

Jon jostled up front. "You idiot, you hit my girlfriend. You could have killed an innocent bystander."

"Just collateral," said the red neck. Jon punched him in the mouth. The red neck reached for his gun but thought better of it. All this was caught on camera. The incident raised questions about gun control, the proper use of firearms and also anger management.

*****

Kate and Jon did not go to any other rocket launches for quite a while. Kate preferred tracking satellites with her telescope, and Jon preferred writing about what she saw for the magazine.

During the TV interview, in the aftermath, Jon was asked why the terrorists might have insisted on canceling the launch and why they felt that Allah did not want man to go into outer space. Jon commented that it was the same old story of ignorance, that the Church fathers themselves had exhibited such stubbornness when Galileo had asked them to look through his telescope at the moons of Jupiter. "Just look for yourselves!" he insisted. They feared to discover things with their own eyes which contradicted the Bible.

"Same old ignorance," said Jon. He figured that if extraterrestrials were found out there, then Islam would have to include extraterrestrials in their conversion tactics, a bigger job than just making Earth into one Caliphate. The interviewer chuckled but then thought better of his reaction, just in case any extremists were watching the TV interview.

*****

When Jon visited Kate in the hospital, after the bullet was taken out, he told her, that the crazy notion that the terrorists had was not so different than his own grandmother had, who refused to believe that man had ever landed on the moon in 1969.

His father recalled, "Your grandmother told me that it was all a hoax, staged by the TV network. God does not want man on the moon. God won't allow it."

Jon could never understand his grandmother's attitude, a clever woman, who spoke 5 languages fluently and had weathered the chaos of World War II and survived that conflict with her wits.

Jon himself remembered his grandmother saying to him during an episode of Star-trek, "You watch that garbage?"

That was one of the reasons Jon became infatuated with astronomy, though his low math marks kept him from going into the field. He earned a respected name among journalists for his ability to write about space. Jon wanted to distance himself as far as he could from his closed-minded ancestors. If he could not breach the secrets of space with an ability in math, he would use his ability with the written word.

22
Meanwhile

Meanwhile, astronomers were eagerly anticipating the launch of the James Webb Space Telescope. It would send back new telemetry to an Earth whose races suffered from never-ending wars and famine.

Some pacifists wondered how far the 10 billion dollars could have gone to stave off world hunger. That didn't matter anyway because the decision was made to launch the craft. James Webb would replace Hubble. It would pacify the hunger for new data from outer space that most astrophysicists craved.

Jon and Kate were just as eager as the rest of the scientific community to have James Webb up there, one million miles away from Earth, sending back sharp pictures and data. We will find out what makes the universe tick! It would happen soon.

Jon and Kate watched commercials on TV which showed slogans, "No Child Hungry", asking for funds to keep kids in America from going to school hungry. What irony!

Kate looked at Jon over her TV tray and commented, "Thank goodness for our good jobs."

Kate and Jon loved their jobs. Kate wrote code for seeking out exoplanets; Jon wrote his articles on what Kate uncovered. His knack was putting her astronomical jargon and numbers into simple terms that the everyday person in the public could understand.

When Kate and Jon were out canoeing, they rarely talked. They understood a private language that only a couple long used to each other could understand. They never raised their voices against each other. They enjoyed their comfortable quietness with each other, whether out camping or at home in their little apartment.

On weekends, during the previous summer, before they were even invited to the conference in Washington, they searched out good lakes for canoeing excursions. They drove out regularly with their RAV4 Toyota, strapping their canoe on top, packing a tent and a cooler with drinks and sandwiches, and off they'd go. They paddled the stress of work away with each stroke.

In winter, they cross country skied, and found the snow and the freshness of the air invigorating. Their work, whether recording the numbers for an exoplanet, or the written word to describe that exoplanet, was likewise invigorating. They enjoyed their quiet little lives.

Before becoming an astrophysicist, Kate thought about becoming a veterinarian. She loved animals. Animals were simple, and except for basic needs, did not ask for much. She had not brought up the subject of marriage, nor children in a long time.

Kate googled the internet for puppies. She found one, a three-quarter Maltese, and one-quarter Shih poo. A tiny little thing with big black eyes. Her fur was white, with light brown on half of her head, and black spots on her paws. She had a button nose. She weighed all of 5 pounds.

Kate and Jon picked her up on a Saturday morning. The little dog scampered around them waging its tail wildly. It barked with excitement. Kate had already decided on her name, "Paisley". Paisley was an ornamental pattern on material.

"That's what she reminds me of," said Kate. The dog was quiet around the house and loved to play, chasing her favourite red ball.

"Simple pleasures, just like keeping a boyfriend," quipped Kate. Jon was uncomfortable with the remark. What subliminal meanings were hidden behind the comment?

No, Kate had not brought up the subject of marriage, nor children for a long time. Though she felt that a pet was a poor substitute for what she felt was missing in her life. Jon was not on the same page.

23
Next Step Stymied

Kate was increasingly unhappy with Jon after 5 years of a relationship which didn't go anywhere. She expected it to evolve; needed it to evolve.

This lack was brought to the forefront of Kate's thoughts especially after she was shot. She thought about her mortality and thought about what she wanted to do with the rest of her life. She wanted a career and also a family. Why not have both! But Jon Hart was in another universe.

She tried bringing her concerns up one weekend. Jon did not want to hear about it.

Kate wrote him a letter.

> Dear Jon:
>
> Since you do not want to discuss the subject of marriage, nor children, I thought I'd express my thoughts in writing.
>
> I want to get married, and I want to have children. I am not happy with our arrangement because I feel unfulfilled. You need to make a decision to set a date and to have a family with me, or we have to go our separate ways. It's been five years, and nothing has happened.
>
> Lovingly, Kate

Jon, like a typical man, gave her the silent treatment. He didn't want to talk about it, period!

By the next weekend, Jon found his stuff packed outside her apartment door. She had changed the locks, and she had moved into a motel for a week or two. Jon was left floundering on his own.

He got angry.

"Alright then," he said to himself, "if that's the way she wants it." He would not see her again.

Anger has a short fuse.

They did, however, see each other again on "friendly terms", if somewhat subliminally "frosty". Kate even gave Jon a lead on a story that explained a new technique in the search for exoplanets. Their relationship had indeed evolved but not in the direction either one of them had wanted.

Soon, Kate was dating someone else at work, and so was he. Kate would eventually bring her life into the direction she needed for her fulfillment as a woman who wanted a family. Jon went his own way. "Like a rogue planet," thought Kate. Was he the rogue, or was she?

24
Incidental

Although Jon put up a good front, he couldn't sleep well at night. Guilty conscience? Life unfulfilled? He missed Kate.

He saw an old high-school friend in whom he used to confide in the old days. They occasionally saw each other now that Kate was gone. The talks they used to have in high school helped them see through many a turbulent time during their adolescence. Now that they were in the working world, they decided to renew their acquaintanceship.

"Getting used to life without Kate?" asked Lee.

"It's getting easier. The word commitment scared me off."

Jon told Lee about his dream. "You know," he said, "I rarely dream, or I rarely remember my dreams."

"I wouldn't worry about it too much," said Lee. He recounted the story about the renowned physicist, Richard Feynman.

"You know, he used to ask people for fun, 'Guess what happened to me today?' His answer was, 'Absolutely nothing!' He wanted to point out to people that when they ascribed significance to a dream, it really meant nothing." [Lawrence Krauss, A Universe from Nothing, p.121]

Lee hoped that this little anecdote would assuage Jon's anxious state. Nevertheless, Jon went ahead anyway to tell his friend the strange dream he had without seeing any message in it. Lee listened without interruption.

I stood at the ticket counter at a movie theatre, frustrated. I had asked for a ticket. My $20 lay inside the portal of his booth untouched. Which movie did I want to go see? I had a momentary lapse and could not remember the name of the movie? There were no posters to assist me. I asked him to list the names of the movies playing. He could not help me; he did not remember the list either. Then the qualifications to get in got more complicated. The ticket man asked me: Why was Schubert's Symphony called The Unfinished? I was in a quandary and the movie had probably already started with my friends inside and me still outside, not even having bought the popcorn yet. I turned around ready to go home. At least, the frustrated ticket man was as ignorant as I was, and he probably would never see the movie either. As to Schubert's Symphony? It only has two movements, and a Symphony is supposed to have four movements. Schubert was absent-minded and put things off. He was a procrastinator. And speaking of being absent-minded, in all the frustration, I forgot my $20 at the counter when I walked away. Some dreams leave you short-changed! It's not a good way to rest when you are supposed to be asleep.

"There you go," said Lee, "just like a dream. Lots of ado about nothing! If there's a lesson to be learned, it's this. Don't pay the man before you get your ticket! And don't forget to pick your money up when you leave! Too bad there were no posters about what was playing!"

They both laughed. Lee was a lawyer now. Jon respected that.

Jon said, "I'm going to tell you something I never told anyone else before."

Jon had always been self-conscious in high school. Maybe that was why they were drawn close to each other in friendship. They both were self-conscious, unsure of themselves.

Jon stirred his coffee, sipped and took a deep breath. He revealed something from his past which he thought would predestine him to a life of failure. Lee kept his mouth shut and let Jon talk.

> When I was in grade 10, we had a math exam. I was not too bad in grade 9, but as grades 10 and 11 came along, my excellence dropped to good and then to satisfactory. By the time grade 12 rolled by, I graduated with a 55% average in math, which was embarrassing.
>
> Students were streamed in those days. The cream of the crop would become engineers and doctors and then, there were those who fell into the Arts. I was in the group which would go into the Arts in university, goodbye Sciences! Science students were smart; that's where the money was, but I liked writing poetry.
>
> Things were already determined back in grade 10 during that math exam. I was struggling with

question #1, when Jim Moyer pulled out his compass and was solving question #3 already. He went on to engineering at the University of Waterloo. What happened to my high hopes back in grade 10? I felt like the bus had passed me by. There were so many avenues open to me then, to all of us, in those days with government grants and funding. You just had to be smart enough and have the marks.

Was it my parents arguing all the time? Or realistically, maybe I wasn't smart enough?

During my session with the guidance counselor in grade 10, he asked me what I wanted to go into. I said, "rocketry". He said, 'You don't have the marks.' 'What about something else?'

I had no idea. I walked out of the office, somewhat dispirited.

Where was my life headed? So many of my classmates knew already where they were going when they got to university. I didn't like mixing mortar and lugging bricks during my summer job with my dad. But I could not think of a refined job for me. What would I take in university? What would lead me to a good job?

Lee chuckled. "And yet, here you are a writer for a top magazine."
Jon also chuckled. "And here you are, a lawyer!"

Lee commented, "Those were good days when everybody could go to university without a mountain of debt to pay. My marks in university got better. I steered myself into Law. I found my niche."

"You were always a good thinker on your feet," said Jon, "and a good talker."

Lee confessed, "You know most of our class was going to become engineers. I was in that group for a time."

Lee recalled odd dreams in high school during this time.

> When Father Kroetch drew a proton on the board with several electrons orbiting around it, I couldn't help thinking of a little solar system. I didn't care if the electron was supposed to have an electric charge. When I fell asleep at night, I dreamt that I saw little beings, tiny sparks of life living on an electron planet. I wondered if the electron was populated by little "light people", and if they worried about jobs like we do.

Jon observed, "Maybe, those dreams were just a way to get your mind off your parents' problems. They were arguing a lot like my parents were in those days. It was a rough stretch for both of us."

"We need to talk some more," said Lee. Jon agreed.

They left a good tip at the table and decided to meet every Friday for lunch.

"Talk is cheap," said Jon. "I'll keep you updated on my dreams."

"It's less expensive than a psychiatrist," joked Lee.

25
James Webb Space Telescope
The "JWST"

The world watched TV as President Biden made his announcement from the White House about the first image sent back from the James Webb Telescope on Monday, July 11. He expressed pride that "America can do big things and see possibilities where no one has gone before." The first image was of a tiny sliver of sky from 5 billion light years back in time showing thousands of tiny galaxies. It was Webb's First Deep Field image of a galaxy cluster which they named SMACS 0723.

Another four pictures would follow the next day, Tuesday, July 12, 2022, at 10:30 a.m.

These new Webb images were taken at different wavelengths in the infrared over the course of 12.5 hours. When Hubble sent back its first pictures back in May 1997, it took weeks to get pictures back.

This is a great achievement for mankind when the Wright brothers made flight possible with their airplane only 119 years ago and when Neil Armstrong set foot on the moon only 53 years ago under rocket power.

Jon and Kate found themselves at the same news conference. They nodded to each other. They were cool in their interaction, aloof but friendly. It was difficult to accept them going their separate ways after such sessions. Other than professionally, they did not see each other. Yet, astronomy still held them in a similar orbit.

They were both eager to see the first pictures that James Webb sent back to Earth from its perch one million miles out in space. Infrared photos do not have space dust to contend with. The images were so much better than what Hubble had produced.

The question was whether the 10-billion-dollar investment in launching such a new telescope was worth it.

No one could argue that the imagination and inspiration which the project sparked for all mankind was worth such an expensive effort. No one could argue that the telemetry and photos that came back from deep space were valuable to physics and science.

This was a special moment for Jon and Kate, who were keenly interested in telemetry and images. It may not have meant anything to starving Africans or countries embroiled in war or to Europeans and North Americans coping with the worst inflation in four decades, having to pay high gas and food prices. What good was the Webb telescope to them?

To Jon and Kate, however, the photos of deep space meant a reprieve from more depressing news about the Russians blowing up yet another city in Ukraine. Climate change was getting repetitive and boring too.

Jon and Kate scrutinized the 5 images that NASA released on July 12th, beautiful pictures in color of the death of a star, and of stellar nurseries where stars were born.

An image of "Cosmic Cliffs" showed the ability of James Webb to peer through cosmic dust with sharp clarity. The telescope sent back pictures of "Stephan's Quintet" displaying amazing galactic mergers. There was a clear image of star formation and how gas in galaxies is being disturbed. NASA was thrilled to get images like this so that astrophysicists could study the evolution of galaxies and also black holes.

Kate especially was interested in the telemetry which the telescope sent back to Earth. Webb's enormous 18 segment mirror and precise instruments captured details from starlight that filtered through the atmosphere of an exoplanet. The light spectrum from a planet 1,150 light-years away revealed the signature of water! This was important in the search for habitable planets in the coming years, even though the trip out there for mankind was impossible.

Meanwhile, on Earth, Japan was mourning the loss of their Prime Minister, Shinzo Abe, who was shot to death last week. Sri Lanka's Prime Minister and President are not faring well either where the economy had collapsed, and thousands were protesting the cost of food, medication and fuel. Britain's Boris Johnson was stepping down due to scandals and his flagrant abuse of Covid protocol. Canada's Healthcare system was a mess. The premiers were meeting in Victoria B.C. wanting to come to an agreement with the federal government about funding. Waiting times and a backlog in surgeries had to be resolved in the Canadian health-care system.

Talks were scheduled between the American and Mexican presidents about what to do with immigrants and the economies of both nations.

In American politics, evidence was being gathered by a committee on former President Donald Trump whether he had ties to extremists who stormed the Capitol last January 6th, 2021.

"NASA strongly rebuked Russia using the International Space Station for political purposes to support its war against Ukraine," NASA said [USA Today, Thursday, July 14th, 2022]. This came in reaction to three cosmonauts aboard the ISS showing a pro-Russian separatist flag while the war was going on between Russia and Ukraine. "(It) is fundamentally inconsistent with the station's primary function among the 15 international participating countries to advance science and develop technology for peaceful purposes."

Putin continued his aggressive incursion into Ukraine hoping to rebuild the territories of the old Soviet Union. Nine people and close to 40 were injured after Russia launched a multiple missile strike on residential buildings in the Eastern Ukrainian city of Khardiv. Khardiv is a Russian speaking city and home to 1.4 million people.

*****

Back in space, a million miles from Earth, James Webb was doing its silent work among the stars making strides in uncovering secrets of the universe.

Kate thought, "I would rather listen to what the stars tell me, than to hear more news about war and politics."

It was like Kate was reaching into outer space to get off this planet mentally and emotionally, hoping to find peace out there among the stars.

26
What's Your Address?

Earth cannot be seen with the naked eye from Kepler 22-b. The Keplerians or whatever they call themselves would consider Earth insignificant. They might be too advanced to give a thought to Earthlings whom they might just look down on as busy little ants.

It's amazing how a telescope can adjust its focus from small to super large, until the small disappears. If an alien asked for our address, we would have to give this sequence:

- Earth
- Solar System
- Third planet from the sun
- Orion Arm
- Milky Way Galaxy
- Local Group
- Virgo Supercluster
- Laniakea Supercluster

We are so easily lost in that sequence. They say that the observable universe is 90 billion lightyears across. But what if the universe keeps on expanding to infinity? Would our address disappear in all that huge expanse? Would we get lost?

In terms of importance and perspective, we are so insignificant in the Cosmos. We hardly rate on a technologically advanced scale. Arthur C. Clarke observed,

"Any sufficiently advanced technology is indistinguishable from divinity." That could be an unsettling thought for religions across the world. What were super beings like at the Type 4, 5, 6 and 7 level of the Kardashev Scale?

The 90 billion lightyears of our own bubble of the universe are an incomprehensible distance. As if that isn't enough, we might be only one universe amidst an infinite number of multiverses. Messages and mail to us could surely be lost in such a Cosmos. Where does that put the address of our little Earth?

> "There are billions of stars in our galaxy, and hundreds of billions of galaxies. Our universe itself is also just one of many."
> *Stephen Hawking*
>
> "I believe we live in a multiverse of universes."
> *Michio Kaku*

The largeness of existence is astounding. It's been said that even if we pictured our address to be on a pale blue dot in our observable universe, we don't have to worry about first contact with any of our neighboring aliens. It does not matter if intelligent alien life existed elsewhere in the universe. We will never find them, and they will never find us. In other words, we are effectively alone in the universe.

*****

That is a comforting thought in one way and disappointing in another. If the flag waving on the ISS last week is any indication, then it is good that we will never contact Keplerians or any other interstellar beings, because flag waving and property disputes could cause interstellar war. They would not want to have anything to do with us! We do not even have peace on Earth yet, if ever?

If there is an advanced race out there, then we on Earth would be a brutish race, not worth their while, and not worth giving a helping hand to for further advancement. We would be too savage to civilize!

27
From Fiction to Fact?

In the fourth century B.C., Anaxarchus, a Greek philosopher of the school of Democritus [theory of atoms], believed that an infinite number of worlds appeared and disappeared all the time. Alexander the Great wept for he had yet to conquer even one world.

Almost two millennia later, in 1686, Bernard le Bovier de Fontenelle, a French author, published <u>Conversations on the Plurality of Worlds</u>. So, the idea of intelligent life existing in multiple worlds was not a new idea! Even more revolutionary was the idea that multiple universes existed, and that life existed on any number of them.

Throughout history, mankind has gazed up at the stars and wondered what they were or what they were made of. It is amazing how accurate clever men of ancient Greece were, that they often had thoughts about planetary systems and life on other worlds, even without the help of telescopes and modern astronomy.

Earth has come a long way in its development, despite the wars and civil wars in its history where nations grappled for more territory and power. We plow ahead in the name of progress while we kill one another.

Maybe scientists and astronomers are a blessed breed if they are funded by government grants for what they love to do. They seem fulfilled. They enjoy a sort of protected bubble for clean and modern labs and equipment for their research, while the rest of humanity lives on a different level and in a different class of people Maybe, the message to our progeny is to be smart enough to belong to that elite class of thinkers and visionaries!

Those who do not belong to this august group should be jealous. The rest of humanity has to toil in factories or in the fields to work for a meagre living, while the smart people of the world stare at the stars and dream.

Maybe that is appropriate and the natural order of things. How it should be. We should be grateful that this upper stratum does exist and is able to explain the stars to the rest of us. Physicists have, after all, given us the light bulb, the computer, the laser and transportation to the moon. And we've also had a wonderful understanding of the stars from such people as astronomers and physicists.

The farthest the Hubble Telescope has seen is the GN-z11 galaxy, which is 13.4 billion light years away, compared to the Big Bang which occurred some 13.7 billion years ago. There is a 300-million-year gap which we have not seen, nor measured.

But that's not quite right. The GN-z11 galaxy is, in fact, 32 billion light years away from Earth, not 13.4 billion, because of the constant expansion of the universe. The limits of outer space are forever out of our reach. Such is Hubble's cosmological constant of an ever-expanding universe.

The James Webb Telescope was built to fill in crucial gaps where the Hubble Telescope fell short. The JWST is twice as large but only half the weight of Hubble. The mirrors are 6.5 meters compared to Hubble at 2.5 meters. We have keener eyes now to peer into the Cosmos and to understand its complexities. Astrophysicists are elated that they get finer telemetry sent back to them, making the 10 billion dollar scope worthwhile after all.

The Keplerians:
TYPE III, a galactic civilization

The exoplanet Kepler 22-b was not eager to be found by any Earth telescopes. Keplerians were alerted to the new eye in the sky that was trying to watch them, the newly improved James Webb Telescope with its almost 22-foot mirror. The machine was launched on Christmas Day, 2022, and stationed one million miles away from Earth, free from obscuring space dust.

Kaal-el made his presentation before the interstellar committee. "We have detected an advanced telescope which has been snooping on our planet," he told them. "These Earthlings, as they call themselves, have sent a telescope into space which can probe into our atmosphere, and perhaps soon, can detect that we are here. This may not be a good thing either for them or for us."

He and his assistant, a female astronomer named Elina, proposed that Kepler 22-b send back erroneous telemetry to the Earth home-world.

"It will keep them off our tracks for some decades. As it is, they are a warring race, and it would be best to keep them away from us."

The Keplerians were a Type 3, a galactic civilization, because they drew their energies from the stars in the galaxy. They had several Dyson spheres which they built around red dwarf stars, so they could run the lights in their cities, their industries and also the power in their starships. They had cities which floated in the clouds, as well as cities that were built under their seas.

They had a functioning global government which ruled for the good of the whole planet, and not the segmented components of the planet known as countries. There was only one flag for Kepler, and it was black with a large white image of a galaxy in the center.

Keplerians were stocky and strong. They had to have strong bodies to withstand their gravity, which was twice that of Earth. They were working on transcending their bodies and had progressed enough to enjoy mental telepathy but only in face-to-face proximity. Distant communication, up to this point, was too muddled and confusing because they had trouble focusing on only one mind. When their ancestors negotiated peace in their world, mental telepathy on a one-to-one basis, at least, verified intentions among politicians.

Kepler was close to a Type 4, a universal civilization, which could create its own wormholes for galactic travel. Their spaceships traversed regions from one end of the galaxy to the other through these wormholes.

Gold and diamonds meant nothing to them, since they could travel to any planet within the Milky Way to retrieve any materials they wanted for their buildings, their machinery, their fuel and their ornaments.

Actually, few of the Keplerian women wore ornaments. They wore their ornaments, as it were, upon their souls. The motto of the Keplerian home-world was, Health is Wealth. No one got sick on their planet. They were disease free. Their bodies simply wore out from centuries of use. But they were working to remedy that too. They knew that in a few other worlds, though they were rare, beings existed who were free of their mortal bodies. They were "spirit".

Keplerians knew that this state was possible. A committee of doctors told Kaal-el, "We're working on it."

When Kaal-el scoured through the history of the ancients, he found that eons ago, it was his race which had seeded the formative life on Earth. In fact, Kepler was responsible for the evolution of life on many planets in the Milky Way.

"They are our children," said Kaal-el to the global committee, "and so we should feel responsible for their growing up. It would be a shame to have them annihilate themselves because they have evolved technology which would allow them to do just that. At some point, we might have to intercede and save them from themselves."

It seemed that when the Keplerians seeded the Earth with an intelligent species, hundreds of thousands of years ago, they had to give them religion to give them principles on how to behave and ethics in terms of boundaries for what they must and must not do.

Their evolution would someday bring them beyond religion and into the stars. "That is when," said Kaal-el, "it will be a good time for them to meet us."

Kaal-el said to his government, "For now, let the Earthlings probe. This James Webb Telescope is a wonderful achievement for them. It can see beyond the dust in space and spy upon us. Once they learn more about us and other worlds, that may be the time to extend our hand in friendship."

*****

A faction of the Keplerians did not agree with Kaal-el. They figured that sending false telemetry back to Earth was the way to go.

"These Earthlings are deceivers from what we've observed of their history over the eons," they said. "They deserve to be deceived in turn."

They justified this as a tactic that was meant for the good of the Earthling species. False telemetry was easily fabricated with the advanced technology that the

Keplerians possessed. Would the Keplerians be able to fool the James Webb Telescope?

How fast would human technology progress until they could not be stopped in their understanding of the stars and of what possible life resided on other planets?

28
What's in a Name?

Space is cut up by people who think they own that bit of real estate when their names get put on it for some kind of discovery. That's a scientist's ticket to immortality.

It's amazing how planets mind their own business, doing their own thing according to the laws of physics, while human beings toil down below on a little plot competing with other experts trying to shout the loudest who has unlocked secrets to the universe.

The planets and stars don't care. They leave themselves open to discovery without any care about whose name is on a new discovery. They cannot be owned, at least not yet, until some civilization builds a Dyson Sphere around some red dwarf star.

Envy is one of the Deadly Sins, if there is such a thing as sin. Intellectual property should be credited and not stolen. People, by nature, are not honest.

Experts are jealous of their territory. They rush out to a publisher to get an idea into print first before anybody else gets there. Often, it's questionable if the idea was original or not. It could have been something they overheard in a conversation, as a speculation, which they take to heart, nurture and then take as their own. Maybe, the idea grows commensurate with their own egos.

Claims about intellectual property can get dirty. Plagiarism crosses many fine and not so fine lines. People feel they have a right to someone else's idea if something was not published. They run with it without giving credit or feeling guilty.

On Earth, competition can be fierce among professional people, both men and women alike. Everybody wants to have their name immortalized, or at least, recognized and appreciated.

Astronomer Sara Seager, writes about a friend, Dave Charbonneau, who had the proof of a full transit of an exoplanet across its sun. Apparently, his discovery leaked out and two colleagues who only had partial proof of a transit scooped him by publishing their material first. [The Smallest Lights p.71]

*****

Sara Seager herself was sort of shuffled under the carpet with her own original idea about the spectrograph of the atmosphere of an exoplanet. It could be that she was a woman and astronomy, even in these times, was a man's domain. Seager's idea was to use the same sunlight which revealed a transiting planet as a method to determine the makeup of that planet's atmosphere. Analyzing its atmosphere through a spectrograph would reveal the atmosphere's gaseous makeup. Biosignatures like oxygen and methane could indicate life. Seagar and her colleague, Dimitar Sasselov, published a paper to that effect, called "transit transmission spectra." [p.74]

What happened was that NASA gave a team, who had cited Seager and Sasselov's paper, the right to study the light filtering through a hot Jupiter. Seager, even though she came up with the original idea, was not included on the team. Seager writes: "My theoretical efforts had triumphed, but I still felt as though I hadn't been invited to the party I had helped to throw." [p.74]

Well, Frank Sinatra sings, "That's Life."

29
Parallel Injustice

Maybe, something similar happened, at a parallel time, in a parallel galaxy far, far away on a parallel Earth.

Proponents of the multiverse theory speculate that identical people might be living parallel lives on some Earth-like planet in another galaxy or in another dimension. We have duplicates! Some planets may be exactly like ours; some planets may be similar but not quite. How amazing is that?

Perhaps, in some other universe, things may turn out more to our liking. Perhaps, a parallel Seager was not shuffled under the carpet. She might have been asked to be on the team by a parallel NASA to study the light filtering through a hot Jupiter.

Perhaps, that universe has given women an even break throughout history. Given that women are our other half of humanity, it only seems right that women have equal rights, as it should be, as it always should have been, everywhere in time and space.

But chances are that even in another Earth-like reality, the same mess exists there too, and all the inequality we've suffered has taken place there too. One can only shake one's head at the unfair Cosmos.

The planet upon which this story happened was named Aerth. Strange, that the names of people were so similar too to our Earth. The story unfolds:

Aerth:
Type 1, a planetary civilization:

Edo Kucher was eager to please Kathleen. Edo was a graduate student whom Kathleen had taken under wing to mentor. Kathleen hoped he could live up to tracking the data on a possible transiting planet across a test star. That was a plum assignment which could have won him many kudos if he would have stuck to his job, but he had a forgetful and careless nature.

He started the work in June but did not look at the data until 3 months later in September. Who knows what he was doing in the meantime? Kayaking or white-water rafting in the Grand Canyon?

Yet, a bundle of data accumulated in his computer at the beginning of September. The data clearly showed the complete transit of an exoplanet. It does not matter now, why he didn't get back to his work for so long. But the consequences were embarrassing and could have been deadly as the following made up story suggested.

The lesson there, if any, is that the student should not have procrastinated so long in his assigned task. That's the price you pay for getting distracted.

When Edo returned in September from whatever he was doing, he glowed with satisfaction that his data showed there was indeed a dark silhouette which crossed the light of the star he was studying.

He wanted to share the news right away. Well, share was not actually the right word. He yielded to a bit of braggadocio, so he bragged to Kathleen about his find. Proof of an exoplanet! Kathleen, in turn, was pleased, and passed the good news on to several other colleagues around the water cooler. The water cooler buzzed with the news.

Edo invited Kathleen and several friends out to the local pub to celebrate the find. Brian Harris said he would stay behind to catch up on some work. He was already a celebrity in the field and apparently felt no compunction to compete.

When the group had vacated the department, Harris entered the grad student's office. Offices were a shared domain among grad students and doors were left unlocked. Harris saw papers on top of Edo's desk, the exact papers he was looking for. The telemetry was all there on computer paper. Proof of a complete transit from start to finish.

Harris and his partner had recorded the same transit but later in the season and it was only a partial eclipse. With Edo's telemetry, Harris had verification of a full transit, the way astronomers verified the existence of an exoplanet. He and Turner only had partial proof; Edo had irrevocable proof.

Harris pulled out his cell phone and took snapshots of the telemetry he needed to back up the proof for the existence of exoplanets.

He and his partner had already written a paper which was incomplete, but complete enough, to make a claim. Their names would be bantered about in the media and in reputable journals.

Harris persuaded his partner Matt Turner to polish their paper enough to publish. They wanted to get published first, while their colleagues were quaffing beer at the pub.

Harris' partner had some misgivings about doing this because the grad-student, Edo, had after all definitively proven their claim. They only had partial proof. Turner wanted to give Edo credit in their paper. Harris said, "No."

"Ignore Edo," said Harris, "he's just a grad student. Ethically, we have every right to publish, even if we've gotten partial proof from a partial eclipse."

There was no way that Harris was going to share a byline with a pipsqueak who hadn't even gotten his Ph.D. yet!

"Why should we give him credit and let his name weaken ours when we were the first to entertain such a theory."

It's not that Harris actually used any of Edo's telemetry, though the temptation was there. After all, he had photographed the data on an impulse.

"The point is we don't need to use it," affirmed Harris, deleting what he had on his cellphone. He challenged his partner to revamp their paper and get the thing out there in print.

Harris and Turner worked all night. Their paper turned out to be a well written article. First, they submitted a brief abstract in the magazine, <u>Astronomy Now</u>! The news release stated that the first transiting planet had indeed been found, and that Harris and Turner were the ones who found it. They had proof, at least partial proof, with more to follow. They also sent their article off to several other reputable journals.

30

Scooped!

Poor Edo was scooped! His name wasn't even mentioned. He had to publish his findings after the fact.

Edo stewed about the situation for days. He suspected that Harris had entered his office and seen the data on the papers on his desk. If any of the raw data appeared in Harris and Turner's paper, there would be a lawsuit.

This was a hard lesson for Edo to learn. Don't trust colleagues. Be the first to publish!

Kathleen, as his mentor, was not only his psychiatrist but also his friend. She advised her young protege to go ahead and publish his own findings anyway, even if that followed hot on the heels of the team who got the glory and credit. Edo was not happy but calmed down enough to collate his data and start writing: <u>Definitive Proof of Exoplanets by the Full Transit Method</u>.

However, he could not avoid running into Harris in the hallway.

He confronted the man.

"I know what you did. Someone saw you coming out of my office the day we went to the pub to celebrate."

"You have no proof of that," said Harris, "just like you cannot take full credit for the discovery of exoplanets."

Harris tried to appeal to the young man.

"A partial was all the proof our team needed to get the glory. Sure, data from a full transit would have been better, but the partial for us would do. Get over it, Edo."

Edo was young and he got angry easily. He stewed some more and gave Harris and his partner, Matthew

Turner, the cold shoulder whenever he passed them in the hallway. Perhaps, it was an adolescent thing to do.

Edo had possession of firearms. He and his father were avid hunters. He knew what he would or rather could do with a gun. Revenge was a dish best served cold.

Edo waited. All sorts of imagination went through his head. He spent more time on the shooting range, than at the end of a telescope these days. He vented his anger through shooting inanimate targets.

Edo thought about the perfect crime. No evidence. No fingerprints.

But what was the sense in that if people did not know why Harris and Turner had been murdered. In his mind, Edo had already done the deed. He determined that the world of science would know what the two thieves had done to a fellow scientist. There was nothing so low as stealing someone else's work.

Harris tried to reason with Edo again.

"Look! We had partial proof and that was good enough to publish. The fact that we published first and got the buzz for our discovery does not negate our contribution. Your findings just confirmed our results. Let's face it, you were tardy in coming out with your data."

When Edo talked with Kathleen, she said, "Harris and Turner are right. They had enough data to publish and in doing so early, got enough hype for their discovery. They were right, they could confirm their partial finding at a later date with more data."

Kathleen advised Edo to cool down. If Harris and Turner had published, even their partial findings, using an iota of Edo's data, then there was a case for a lawsuit. But they went ahead with their own data, even though it was partial, and so were not culpable of any breach of ethics. Edo couldn't even prove that Harris had been in his office.

"Look!" said Kathleen, close to exasperation, "forget about Harris and Turner. And forget about anything else you might have had on your mind in getting even. Listen to me. Gather your data together in a well-documented paper proving the existence of exoplanets. Publish it as confirmation of what was only partially proven to date. Don't even mention Harris and Turner by name. It's your name that will stand, in the annals of exoplanet discovery, while they will be forgotten with their partial data."

As his mentor talked, Edo's anger dissipated like gases in a disappearing star.

"Keep the name of astronomy unsullied," added Kathleen. She brought the situation right out into the open, adding, "Can you imagine the scandal you'd have caused, if you had lost your head and shot them dead. That would have done irreparable damage to the good name of science, and it would have sadly ended three careers, Harris', Turner's and sadly yours...because I know you are better than that."

As Edo walked out of Kathleen's office, she said, "There are as many fields of study out there among the stars, as there are stars. Fame is not all it's cracked up to be, although it is nice to get credit."

31
Jon and Kathleen:

Aerth followed the history of Earth almost completely, but not quite in another slice of the multiverse.

Kathleen and another version of Jon were still an item. They went to Cocoa Beach to watch a space launch, not a major thing, just another communications satellite going into orbit.

A fanatical arm of a religious group, the Izzins, was convinced that space was not where God wanted man to be. That was a divine domain. They needed to make a public statement.

They were armed and ready to take hostages demanding that the launch be cancelled. The General in charge was a wise and shrewd man. "We can always reschedule and launch another time."

The SWAT team moved in and the Izzins surrendered having made their point. However, a red neck "cowboy" drew his pistol and shot one of Jihadists dead. There was more gunfire, and another volley of shots went off from the direction of the cowboy. He killed another Izzin, but two of his shots went wild. They hit Kathleen in the chest. She was beyond saving. What a loss, an astronomer who studied the stars and was so involved in space exploration, with all that education and enthusiasm...lost.

Jon found the cowboy and attacked the man. He almost drew his gun on Jon. If Jon had been killed by the same man, that would have been a fine scandal for gun advocates. People could understand Jon's position, especially after news reporters made the relationship between Jon and Kathleen clear.

The incident fueled the campaign for gun control, something the cowboy was obviously against. He spoke to the media, "What we need is more self-protection, and I mean with guns." He insisted, "Who is going to protect you against these foreigners and their weird religions?" He considered Kathleen's death as accidental and purely collateral damage. "Just one of those things."

Jon didn't see it that way. "You need target practice," he accused. "Where do you think stray bullets go?"

Soon, Kathleen's death went the way of old news. Astronomers were looking forward to the James Webb Space Telescope being launched a million miles away from Aerth. They were somewhat behind schedule of the other Earth, but amazingly, they were in line with the technology which would bring them closer to discovering aliens. To many scientists, the real discovery wouldn't really matter. What mattered was the journey...and whether the funding would keep on coming.

*****

Kathleen's ashes were gathered in an urn and given to Jon, as her will prescribed. He took her ashes out on a canoe ride around one of their favourite lakes where he sprinkled Kathleen's remnants into the sparkling water. The sun shone brightly over the lake and the ripples sparkled like so many stars.

Jon regretted never having committed to Kathleen. He knew she wanted to get married, but he had such big fears of what commitment would demand of him.

All he said, as he spread her ashes into the silver sparkles of the water was, "I'm sorry."

Kathleen's dreams of a family and having children were forever unrealized in this universe.

Jon never got married, although he sunk into a pattern of comfortable arrangements with female co-workers in future years. This did not matter to the stars who counted their lives in eons, and barely blinked at the transitory existence of sentient beings. Kathleen was forgotten.

32
Erth

They say that 52 duplicates of Earth are possible. Of course, that possibility existed only in the DC Comic Multiverse. Some as complete duplicates; some as close approximations; some with big differences. Who knows what is possible in real infinite space and time?

Maybe, we will stick with this version of Erth, whichever is the happier one for our cast of characters, especially for Kate and Jon, who in the other two universes were unhappy people. Maybe, they would have better luck in this dimension and this Erth? Maybe Jon is less of a paranoid and more committed to Kate and maybe, they actually do get married.

Jon, like his mother, was always thinking that every silver cloud had a black lining. He just saw black in everything.

"Imagine what we've achieved," Kate said to him, looking at the James Webb rocket that was about to take off into space. He replied, "It's sad when you think about it, that in 5 billion years, the sun will swallow up any evidence that we humans ever existed!"

"At least," said Kate, "we won't be around in 5 billion years to witness that, nor will our kids."

"Did you say, kids?" observed Jon.

"Of course," said Kate. "It's only the natural order of things. We will have kids, won't we?" she asked.

Jon felt it was a loaded question. He merely said, "Yes, dear."

As to the natural order of things, Kate soon sported a ring on her finger, a wedding ring and so did Jon. She felt a lot better about their 5-year relationship now, and so did Jon. He felt better because she felt better.

It was not a gaudy ring, but a simple band, simple rose-gold. That's all she wanted. It was a life-long marriage that counted. She imagined their souls taking off on the rocket, like it was their honeymoon, a million miles away from Erth. She was happy.

Kate and Jon had privileged seating at the Arianespace's ELA-3 launch complex at Europe's Spaceport located near Kourou, French Guiana. They were ready to watch the James Webb Space Telescope be launched into outer space, where it would find a comfortable position one million miles from Erth. The telescope would not have to cope with space dust, and it would have infrared capability to pierce into the deepest reaches where galaxies swirled and sometimes collided. Erth now had a ringside seat of the Music of the Spheres.

Jon and Kate could hardly wait for the first photos from James Webb to come back to the planet. Their thoughts wandered. Their minds had flashbacks about important instances in their lives, and moments of future hopes and dreams. It's amazing what the mind does when adrenalin pumps through the brain.

Kate reached out to grasp Jon's hand as the countdown proceeded. Everything was "nominal" which meant it was a "go". She felt Jon's wedding ring.

Before James Webb, there was Hubble. Hubble was launched in April 1990. Astronomer, Bob Williams, had a crack at the new telescope for a 10-day stint. His colleagues asked him which part of space he would point to with the powerful telescope. He said, Ursa Major. Now, Ursa Major was a blank, black, empty part of space.

"There's nothing there," said his fellow astronomers. "It will be a wasted effort for valuable telescope time." Williams persisted. His persistence paid off. The Hubble Deep Field image came back revealing 3,000 galaxies in the distant void. [The Smallest Lights, Seager p.115]

Maybe, the same incident happened on Earth, the parallel Erth. This time a James Webb Deep Space Telescope looked back into that dark patch of nothing which was supposed to be us in another Ursa Major. We are one of those 3,000 galaxies where a tiny pale blue dot floats in black space. We have two James Webbs, maybe 52 of them, maybe thousands of them, looking back at each other, mirror images spying on each other. What an amazing reality that would be!

But even the reality of such a Cosmos was nothing compared to the reality of their marriage. How did that miracle happen?

Jon felt, for the longest time, that mere friendship was comfortable enough for him; he was not sure about anything deeper.

When Kate, being a bit more open and aggressive, broached the subject of marriage, Jon put up shields.

One evening during a walk under the stars, he confessed.

"Look Kate," he said, "maybe we are not suited to each other." He hesitated and then explained how he felt inferior to her. That he only had a 55% in Math in grade 12, and that his job, as a "wordsmith", was not that lucrative. "I feel inadequate marrying a woman who is my superior, in everything."

Kate confessed, "So, that's what's been bothering you. I don't care about that. It's you I like. Will you marry me?"

Jon was somewhat taken aback. But he liked it.

"Sure," he said, somewhat abashed. It was settled. They had talked about it. She would be the breadwinner if her job paid more. He would be the writer from home, and in his spare time, be the cook and bottlewasher too.

Shortly after Jon and Kate adopted the little dog, Paisley, Kate asked Jon, "Do you enjoy that little one?", pointing to Paisley.

"Goes without saying," said Jon.

"Well, we've got another little one coming our way," commented Kate, patting her stomach.

Jon swept Kate into his arms. "I couldn't be happier...but let's not name him or her Paisley, okay?"

They named their twins, Brad and Brit. She liked the sound of it. Jon was confident that their little Brit would be every bit as smart as their mother and that their little Brad would become a great writer, just like their dad. Life was good.

In the meantime, the couple shook themselves out of their reverie. The thunder of the booster rocket roared. James Webb was launched into space on that Christmas Day. It would be the birth of a new era for humanity. Jon and Kate held hands as they watched the launch.

What would the images of their James Webb Deep Space Field be like?

33
Parallel Experiences

The genesis of this book came out of reading, <u>The Smallest Lights in the Universe: A Memoir,</u> by Sara Seager. There were so many parallel experiences there in her memoir in terms of losing someone and also my own personal interest in outer space, that I combined my own experiences and research into writing this book.

My brother-in-law, Ken Janzen, died on January 18th, 2008, of lung cancer. He took 9 months to die and basically withered away until he was under 100 pounds. Sara Seager's husband, Mike, died of pancreatic cancer. Kenny was a computer nerd, a web designer and sadly a heavy smoker. Sara Seager's husband was a canoeist, an athlete, and a general handy man.

When Kenny was still alive, we would visit, and his 2-year-old son would always say, "Daddy, sleeping."

Sara's family and my brother-in-law, Kenny's, had two young kids and that was one of the reasons that the man of the house did not want to let go, because he felt duty bound as "man of the house."

My job, as Uncle John, with Kenny's sickness, was to take 5-year-old Darriane out for a walk when the family wanted to talk about serious stuff. I could connect in Seager's book when she asked Uncle Dan to take her kids to the park when Mike died.

Sara Seager arranged for a hospital bed to be brought into the house when Mike lost his mobility. Our family had a hospital bed brought into Kenny's house in the basement where he spent his last days. We all took turns watching him until the nurse brought in the oxygen monitor, whose meter read 20%, 10% and eventually 0% after several days.

Kenny created a Wellness Blog where friends and relatives could leave thoughts of encouragement for him. Then, all of a sudden, we got spam, links to porno sites which we had trouble getting rid of. How could someone be so cruel? Someone said it wasn't done by a cruel person but by some automated "program", whatever that means. Anyway, we experienced a spectrum of outrage and family suffering.

Ken Janzen, died at the age of 37. Sara Seager's husband, Mike Wevrick, died at the age of 47.

Originally, I was going to make Jon in this book go through cancer as part of my story in We Are Not Alone, but then I thought better of it.

I'd already written about cancer with my brother-in-law several years ago in a trilogy, called Time in a Bottle, publ. by Amazon, in 2019. Then reading Sara Seager's book, I decided not to enter that dark area of thought again. So, I decided to skip making my imagined character, Jon, go through such suffering.

I did, however, see so many pertinent parallels between what our family went through in real life and what Sara Seager describes in her open and honest memoir. When her son, Max said, "You know, Mom, it's better to have a dad who is dead than a dad who is sick," [p.161] that really hit home.

When Kenny was sick, he could not be a dad, nor do anything with his little kids. Also, the lives of family members became fractured depending upon whose turn it was to volunteer to help. I hate to say this, but it was a relief for all of us when Kenny finally let go. We didn't want to see him go; yet somehow, there was a burden that was lifted.

Rehashing all the details is not good for me, nor does it serve a purpose in this book. Though I must take one more swipe at Canada's failing Healthcare system, which started out proud and universal. Kenny went in circles with misdiagnoses and had long waiting times. That is one of the reasons that I do not want to bring back angry memories in this book over a failing Healthcare system in Canada.

We could have used the intervention of God, or an alien to fly in and save Kenny. But that never happened. Good doctors and prompt help would have helped greatly.

I've done a lot of reading on astronomy, using the opinions of scientists like Brian Greene, Michio Kaku, Lawrence Krauss, David Kipping, and of course, Sara Seager.

It's amazing where we are in this search into outer space. We've only just begun!

As a former Catholic, now a Mennonite, and a Christian, I find some of my journey and research changing my notion of God, which is not what my old Catholic upbringing taught me.

My mental image of God is different and certainly bigger. But these might be questions I might broach in a later chapter.

In the meantime, I enjoy letting my imagination wander and weave little anecdotes of imaginary lives of alien life in other worlds. I hope you enjoy all these possibilities.

Erth is where my consciousness has settled, at least for a brief time, because, as I've said, who wants to rehash the pain at the end of someone's life?

For the moment, Jon and Kate were happy. They were successful in their fields; he in writing for scientific journals and magazines; she in her search for exoplanets and possibly alien life.

Two decades slipped by so quickly. She still has not found alien life, but the search goes on and she is happy doing this. But things change and things come to an end. This applies to personal life and also events in the Cosmos.

The death of a star is a dramatic event. It's beautiful. While most stars quietly fade away, the super-giants destroy themselves in a huge explosion, called a supernova. The death of massive stars can trigger the birth of other stars. The Cosmos is caught up in the cycle of death and rebirth. Maybe, the Big Bang went in a circle, first a big bang, then atrophy, sucking in on itself, then another Big Bang. Both the micro and the macro systems of existence go through natural cycles.

The James Webb image in July 2022, captured a star's final performance in the last throws of its existence. James Webb photographed the gas and dust ejected from dying stars in the Southern Ring Nebula, approximately 2,500 light-years away. The death of a dying star is beautiful!

Not so, when a person dies. Jon had died at 60 from lung cancer, weighing less than 100 pounds. He died peacefully in a drug-induced coma, with family and friends surrounding the hospital bed installed in the basement of his house. He wanted to die at home, not in a hospital.

At least, he and Kate had a life. Their twins grew up to be clever and successful too, following in their parents' footsteps, Brad became a great writer and Brit, an astronomer like her mother. So, the cycle continued. Big Crunch, Big Bang, Big Crunch, Big Bang "ad infinitum".

*****

But now I've written myself into a corner. I'm at a quandary where to go from here? Maybe, I need a couple of sleeps to redirect.

34
What If?

In science fiction, we can explore the "what if" of alien life. What if we could reach them? What if they could reach us? What if they are just like us? What if they are nothing like us? Is alien life something that we will recognize as life in the first place? Everything on earth is something similar, carbon based. What would something from a different evolutionary tree even look like?

These were speculations from science shows on "YouTube": *The First Aliens / A Brief History of Aliens in Science Fiction.*

*****

There could be as many as 40 billion planets in the Milky Way that could sustain life. We've notched up only 5,000 or so exoplanets, still a long way to go!

Michio Kaku says that there could be interstellar highways with aliens zipping around using those highways without us not even knowing about it because we are too stupid to detect them.

Nikolai Kardashev posited in 1964 that there have been civilizations billions or hundreds of billions of years old, that they've already been and are gone, that some will develop yet in future years, that some are baby civilizations like us, and others are mind-bogglingly advanced civilizations, using energy from a star or even a galaxy!

We are a little speck of intelligence in cosmic evolution. We are only at level 0.72 on the Kardashev Scale which has now been redefined by subsequent astrophysicists ranging from Type 0 to 7. We are almost three quarters of the way through Type 0! Not even Level 1 yet!

We assume from our worldview that aliens will look like us. Leave it to the imaginations of creative writers to open our eyes as to what other lifeforms would look like.

*****

I love movies that bring out some of the old scary stories. <u>War of the Worlds</u> comes to mind with the tentacle arms. Then, what my mother criticized me for when I was in grade 11, watching a 1966 <u>Star Trek</u> episode with pointy eared Mr. Spock.

"You watch that garbage?"

Maybe mom was uncomfortable with Mr. Spock's ears. The run-down of species that the series created was amazing, though in my mind they still looked too human, except for the Horta, the stone eating creatures, and the Tribbles, which are cuddly furry herbivores. Let's look at a few other imagined aliens in review:

- Vulcans, who have pointy ears and use logic.
- Romulans, an offshoot of the Vulcan culture. They are more warlike.
- The Borg, who are cybernetic robot like, who assimilate new technology. "Resistance is futile."
- The Klingons, who have ridged craniums; they are a warrior species who seek glory, their heaven is Stovokor, like the Viking Valhalla
- The Gorn, who look like a dinosaur,

- The Cardasians, who look reptilian and are the bad guys.
- The Ferengi, who are small with big ears, sharp teeth and think only about money and profit.
- The Betazoid, telepathic species, Deanna Troi, from Star Trek: The Next Generation,
- Andorians, who are blue and have antennae,
- The Changelings, like Odo who looks like a man but can change into a puddle.
- The Greek Gods, who have superpowers,
- Talosians, big headed aliens who control minds and perception.
- El-Aurians, played by Whoopi Goldberg, a long-lived race,
- The Xindi, some look like bugs, some like reptiles, some like fish...
- The Bajorans, from Deep Space Nine, who have a ridged nose.
- "Q" and the Q Continuum, who take on any form they wish, who come from another dimension, who can play with time and space.

*****

Before I forget, I need to add my latest TV obsession, *Farscape*, where astronaut, John Crichton, discovers a parallel universe flying accidentally through a wormhole, and of course, finding wonderful alien life there, enhanced by the magic of Jim Henson muppets:

- Aeryn Sun, the Sebacean. She looks most like a human.

- Ka D'Argo, the Luxan warrior, who looks like he's got tentacles sticking out of the back of his head.
- Zhaan, a Delvian, who is blue and actually a human looking plant life.
- Dominar Rygel XVI, a Hynerian, who looks like a small frog-llike muppet, very selfish.
- Chiana, a Nebari, who is all grey with white hair and black lips, human looking.
- Pilot, ika a "Servicer", who looks like a mushroom with a big head, used to pilot a living ship.
- Moya, the living ship,

*****

C.S. Lewis made an admirable attempt at imagining alien life-forms in his book, <u>Out of the Silent Planet,</u> published in 1938. This was 26 years before Nikolai Kardashev came up with his Kardashev Scale in 1964 of different levels of extraterrestrials.

Lewis even imagined beings who were super-human and in fact pure spirits. If there were a creator level of beings, that concept reduces our notion of God, because then there might not be a need for God in the Christian sense.

Anyway, in Lewis' book, a mad scientist, Weston, and his crony, Devine, kidnap Ransom, a philologist, and fly their spaceship to Malacandra where they intend him to be a sacrifice for the natives. Ransom escapes and runs into a strange creature, which looks like a seal, called a *hross*. The trilogy is not the easiest to read, but if you can acclimatize yourself to an erudite style, the story is interesting. C.S. Lewis, after all, taught English Literature at both Oxford and Cambridge Universities and knew J.R.R. Tolkien personally [the creator of <u>Lord of the Rings</u>], so the writing style involves a myriad of cultivated writing, very descriptive!

Anyway, other life-forms in Lewis' imagined world are:
- Hnakra, evil ferocious animal, the hross hunt them.
- Pfifitriggi, a species of hnau, with tapir-like heads, frog-like bodies
- Seroni, thin, 15-foot humanoids, coats of pale feathers and seven-fingered hands, they are white in the mountains and red in the deserts.
- Eldila super-human spirits, similar to angels, who regard Deep Heaven as home.
- Maleldil the Supreme Being, the Creator,
- Oyarsa a being, like an angel, assigned to guard a world, they communicate in "Old Solar" speech.

There you go! Aliens and their Variations, in sci-fi literature! All imagined, but then who knows what else is out there in the real Cosmos? Something perhaps stranger than fiction?

35
What You Wish For

Yes, be careful what you wish for!

SETI, the Search for Extraterrestrial Intelligence, is intent on sending out messages to outer space and broadcasting where we are. They want life out there to contact us. They also assume that extraterrestrial life is benign, advanced and willing to communicate back to us. But is that necessarily desirable? Or safe?

C.S. Lewis wrote three books about outer space in <u>The Space Trilogy</u>, where he imagined that every planet had a spirit guardian, an Oyarsa, or an eldila. The only planet which was silent was Earth because the Oyarsa on it was Satan. The Oyarsa of other planets had a network and communicated with each other. Earth was left out. Its spirit had gone bad and was bent. The Earth was known as "The Silent Planet".

We cannot assume that all extraterrestrials are good and that they are willing to help us or share their advanced technology as a kind gesture. The third book, <u>That Hideous Strength</u>, published in 1945, is a story which illustrates the idea of danger from outer space where evil spirits offer strange and nightmarish "gifts". [p. 512]

Some humans may be gullible and want what an advanced race has to offer, but is it good or even right? Do we want our world to be transformed for what a powerful private cult may consider advanced? The spirits in C.S. Lewis' book offer a way to "terraform" the Earth and to change it into what most sane people would consider a nightmare.

The young husband, Mark, in the book, is drawn in by the evil side which is bent on world domination. The wife, Jane, who has terrible dreams, is on the good side which wants to stop the nightmare.

Mark is seduced by the idea of progress where the Institute, NICE, i.e., National Institute of Co-ordinated Experiments, wants to purify the planet with a powerful global government that plans to end all organic life on Earth because it's messy. They draw Mark in by a promise of a promotion, a better job with more pay, but they never quite commit to telling him exactly what his job is.

Later we find out that the private group wants to dominate the world as a Master race. NICE even establishes an Institutional Police and arranges a riot in which potential dissenters are arrested. Written in 1945, the story is reminiscent of Hitler's Germany and the Nazis. NICE wants to gain control of the media and of governmental and social structures in England as a step toward spreading totalitarianism over the Earth.

The lower common people are to be controlled and eventually eradicated. The Master race wants to keep their intelligence artificially alive after the organic body has died, "a miracle of biochemistry." Evil spirits provide the Master race with the technology which will change the world.

All trees are to be replaced with artificial trees which do not shed leaves. There will be no more birds to make a mess. A world without organisms is hygienic. There is no need for being born, breeding or dying. A New Man will be born out of this, free of Nature.

The advanced race has figured out how the brain can survive artificially without a body and live forever. The brain will rule the universe.

The whole proposal to the sane person is actually a nightmare made into reality, unless you are brainwashed. Mark does not see this. He is strung along with the enticement to get a promotion, a better job, with better pay. But, as mentioned, they never tell him what his specific job is.

The group is a secret society, actually a cult. Mark says he does not believe in God, but the group's intention is to create a new God in a new future.

They promise Mark the privilege of meeting their Head, which is to be taken literally, a guillotined head that is kept alive by tubes. A Head that speaks and gives orders and yet is not conscious in its own identity. It is a nightmare come true, which Mark's wife, Jane, dreams about and is horrified over. The evil group which obeys the Head considers this as the next step in evolution. [p.532] The human body is considered redundant. There is a plan to create a Technocracy, with a possible world of living heads who would be the Master Race. [p. 594]

However, at the moment, there is only one Head, which has no soul, is alive, and yet has no life of its own. It is a communications mechanism through which the Macrobes speak and run the secret society of evil men.

Macrobes are bodiless lifeforms which inhabit the space between planets. They plot to eradicate mankind and take over the world, all under the guise of advancing society into a new and enlightened order. They have already brainwashed and possessed a cult of gullible men to do their bidding.

The two world wars were a part of a program to pare down what was not needed, all in a plan to cull the herd, to get rid of the riffraff from the human race. The Macrobes would rule the world through their mouthpiece, the Head.

This is the type of horror that C.S. Lewis warns us humans about, so we will not invite the evil powers in outer space into our world.

H.G. Wells is another author who wrote <u>War of the Worlds</u>, published way back in 1897. The story is about invaders from outer space who want to eradicate the human race.

The message? We are not alone, and we are not safe in our private little corner of the galaxy.

36
The Space Trilogy

C.S. Lewis, a Christian, wrote an honest apologia for his belief in God and Jesus in, <u>Mere Christianity</u>. Yet, <u>The Space Trilogy</u>, published between 1938-1945, weaves a clever sci-fi story with a hierarchy of beings right up to spiritual beings including the highest order which is God-like. This story, written before and during the years of World War II, predates Nikolai Kardashev's prediction of different levels of civilizations on other planets.

Within Kardashev's view of existence, if God exists, then he is an evolved product of the natural order of things. C.S. Lewis in his outer space trilogy imagines a similar hierarchy.

One can argue that if civilizations of different magnitudes exist on different planets, then there is no need for a "real God" outside of time and space. Atheists would be right, after all.

There are some interesting speculations from C.S. Lewis which apply to man's exploration and exploitation of outer space. Privately, one assumes that C.S. Lewis himself does not hold this notion, however, that there is no real God. He, after all, was a Christian. But his imagined story falls in line with the Kardashev concept of evolution right up to the level of the spirit world.

It is interesting that Mormons apparently believe in the evolution to godhood. If a person receives "exaltation", they inherit all the attributes of God the Father, including godhood. Mormons believe that these people will become gods and goddesses in the afterlife, and will have "all power, glory, dominion, and knowledge".

One can translate this in terms of string theory, that "Heaven" is folded into another dimension after one has died where one becomes a spirit after the "mortal coil" is left behind.

Maybe, there is a more natural way to get there by slow evolution where humans will eventually evolve into spirits within a million years through a natural process of leaving one's physical body behind. C.S. Lewis speculates in, That Hideous Strength, that "evolution means species getting less and less like one another. Minds get more and more spiritual, matter more and more material." [p. 619] This is all amusing speculation, but possibly possible, if one seriously entertains the idea of 11 dimensions and a multiverse within the Cosmos.

*****

In C.S. Lewis' sci-fi story, Out of the Silent Planet, publ. 1938, Weston, the scientist, is the imperialist who wants to colonize outer space. Ransom is the protagonist who considers Weston's philosophy of imperialism and colonization of other parts of the universe as "raving lunacy". This was written at a time when Hitler was making plans to goosestep his way across Europe, with England in his sights.

Weston, however, sees himself justified in his imperialism, by the survival of mankind, so that mankind would be able "to crawl about a few centuries longer" in some part of the universe. [p.24]

The *sorn* are a type of hnau [being] who have a rule of law governed by wisdom and a sense of hierarchy. Beasts must be ruled by hnau and hnau by eldila and eldila by Maleldil, i.e., beasts by humans, humans by spirits and spirits by God. [p.94]

One of the "pups" or pupils says that the human world is bent because it has no Oyarsa or God.

The Earth or Thulcandra is indeed bent because it has no spiritual guide, no God. If we encountered God, we would be afraid, as Ransom confesses.

"What are you afraid of, Ransom, of Thulcandra?"

"Of you, Oyarsa, because you are unlike me, and I cannot see you." [p.109]

We have to remember that this book was written before the Second World War, and before NASA and the Search for Extraterrestrial Intelligence existed. Therefore, Lewis understandably puts words in the God's mouth: "Thulcandra is the world we do not know. It alone is outside the heaven, and no message comes from it." [p.110] Planets each have a guardian angel, but Earth alone seems to be abandoned and silent.

This is where Oyarsa recounts how the Earth got abandoned, that there was a war in Heaven between good angels and bad angels.

"That was before any life came to your world. Those were the Bent Years. [p.110]

Oyarsa recounts how the Oyarsa of Thulcandra became bent. "There was a great war, and we drove him back out of the heavens and bound him in the air of his own world as Maleldil taught us."

Maleldil is the God of gods.

Here the story exposes the reason why the scientist, Weston, wants to colonize Malacandra, namely Mars. He wants to collect the "sun's blood" which washes down from the mountains. It is gold.

Oyarsa regards such hnau as stupid and also bent. He would not allow humans to load up their spaceship with any more of sun's blood until one of their races comes to speak with Oyarsa.

Ransom confesses the evil intention that Weston and his world would have by destroying the people of Malacandra, "to make room for our people." [p.112] Ransom explains that Weston wants to jump from world to world plundering them all for their riches. Oyarsa asks Ransom if Weston is, "wounded in the brain?"

It's amazing that Ransom confesses this evil against Oyarsa's world and that he openly offers for all three of the interlopers to be killed, including himself.

When Weston and Devine are captured and brought before Oyarsa, the foolhardiness and bombast of colonizers is demonstrated. Weston has killed a few of the natives with his rifle. He talks down to the natives and even to Oyarsa, "we talkee, talkee." He threatens to blow them all up.

All the beings break out in a roar. Weston takes this as a way to frighten him. He says he will not be frightened by this roar. It is not a roar, says Oyarsa. "They are only laughing."[p.116]

This communication is miscommunication. Lewis does a good job in his story in how different species can easily misunderstand each other. He also underscores the fact that humans tend to underestimate the intelligence of other beings, taking for granted that "we" are superior.

Lewis describes the motives of humanity for spreading out to use up and spoil other worlds. Weston would kill all the people of Malacandra, replace them all with Earthlings after something had gone wrong with his own world, "and then if something went wrong with Malacandra, they might go and kill all the hnau in another world. And then another—and so they would never die out." [p.114]

*****

Obvious points have been made in Lewis' space trilogy about the hazards of First Contact.

But more than that. If one were an Atheist, who at least believed in higher orders of evolution, then speculations could run rampant over superhuman events in the history of Holy texts.

God's war with Satan, for example, is a questionable event which demeans the idea of a "real" God who lowers himself to fighting with half of his angels, even if he sits back as a non-participatory General. The concept of God in a war is not God-like. God's creation of Adam and Eve and the Garden of Eden is also questionable about an all-knowing God for creating fallible beings. This makes God fallible.

As to the infallibility of the Bible, many incidences in the Bible could be explained by the existence of supernatural beings from another dimension, as speculated in the higher order of the Kardashev Scale.

Moses' people cross the Red Sea by the power of "force shields" from the spirits in a higher order. Could the fiery chariot that carried Elijah to heaven have been a rocket ship? Could the angel who announced that Mary would bear a child be a spirit from another dimension able to tell the future? The spirit that hovers over Jesus at his baptism like a dove could also be a being from another dimension, falling in line with the Kardashev Scale. Jesus' resurrection and ascension could be made possible by an evolved superpower, God, from Type 7 of the Kardashev Scale. Muhammad's trance when the angel Gabriel talks to him, could also be an intervention from a super race of spirits. All these Miracles are possibly explained by extraterrestrial powers.

I do not subscribe to these speculations, entertaining as they are.

But in terms of the Kardashev Scale of different levels of energy and civilizations, these things can be explained in terms of good sci-fi stories. Let someone else, with the mentality of Frank Herbert weave sci-fi stories around such notions.

37
What Forms?

If extraterrestrial life is found someday, then it may be discovered in three forms:
1. Microbial
2. Plant and Animal
3. And finally self-aware, intelligent life taking on an unspecified form.

Microbials are most likely the first form we will find. It will make big news on Earth. If we hit a jackpot, it will be a combination of microbial, plant and animal life. The real jackpot would come by discovering intelligent life. Here we will have to tread carefully and diplomatically. Their interests may not be our interests!

Many imagined species have been made to look too humanoid. But maybe, that is what alien life would have to look like, humanoid! It could be a natural evolution. A head, a brain, two eyes, nostrils, a mouth, a neck, a torso, two arms, two legs, a digestive system.

Even on Earth, other species mostly have such bodily descriptions and functions. Even a snake has a head, two eyes, two nostrils, a mouth, a lickety tongue and stomach.

I suppose monkeys went through the stages of Neanderthal, Cro-Magnon, Home Sapiens and finally Human, as a natural direction of development. Maybe, that is the progress which life had to make if it wanted to use tools and eventually build civilizations. Professor Michio Kaku says that mankind became civilized because of three things: the opposable thumb, eyesight in the front, and language. He is convinced that this is how our intelligence evolved.

He figures that manipulating the genome of chimpanzees might even give them higher intelligence and language. Check YouTube: "Michio Kaku on the Evolution of Intelligence".

It is interesting what he asks next. If you manipulate a chimp to become more intelligent, what do you get? You get a chimp that's more human. We already have humans, so why bother!

Author Adrian Tchaikovsky asks in a 2015 novel, what if they look like spiders? This gets away from the anthropomorphic image of making an intelligent species in the shape of humans. Tchaikovsky writes about a genius/mad scientist, Doctor Avrana Kern, who has developed an intelligence virus. She intends to save humans by applying this virus to monkeys for the survival of our species, after the human race has left a dying Earth. There is a terraformed Earth where she hopes mankind will evolve and be replanted. However, the intelligence virus inadvertently does not go to monkeys but to spiders. Two civilizations are on a collision course!

Check YouTube: "Children of Time: How Far Does Evolution Go?"

*****

We must think outside the box. What about other life forms, based on silicone, for example, not carbon?

Living organisms must be built from six elemental ingredients, according to biologists: carbon, hydrogen, nitrogen, oxygen, phosphorus and sulfur. [CHOPS for short].

What if the base element was switched around a bit, depending upon the atmosphere and make-up of the planet? A bit of evolutionary juggling!

The recipe to create a living organism, according to biochemist, Matthew Pasek, is: "With a few exceptions, is CHNOPS, plus a dash of salt and a few metals." So, take these ingredients, mix them with a primordial soup and an atmosphere, and you should come up with a bigger framework for life and how varied lifeforms can be!

38
Meaning in the Universe

Brian Cox, physicist, was asked by Joe Rogan, "Are we the only intelligent life in the Universe?" Cox liked to think in terms of our own Galaxy. "At the moment I like to think that there's only one civilization in the Milky Way, and that's us."

Why is that important? Cox says that this gives the context of existence "meaning", certainly or him, even how we behave politically because it gives life value. Since we are fragile beings, "shouldn't we consider ourselves to be extremely valuable?"

Without us, there would be nowhere else in the Milky Way, where meaning would exist. Richard Feynman speaks about the same thing in an essay called, "The Value of Science." Cox considers Feynman one of his heroes. He says, "What is self-evidently true is that meaning exists here because it means something to us."

If it wasn't for us, then there would be no meaning. We are a very rare configuration of atoms, says Cox, so we are "the only island of meaning in the Galaxy."

Cox shies away from a God concept. He says that humans need to then assume the responsibility of meaning which to him is a grander concept than if we were created by God.

Of course, Cox concedes that there are probably other civilizations out there in the galaxy, merely by the existence of two trillion galaxies. He thinks that it's so rare that there could only be one or two civilizations per galaxy. Still the idea of such a reality is astounding. There could be civilizations out there that are far more advanced of us. Why? Imagine what they'd be like if they survived, 100 million years, 1 billion years? Within 500 years, we've walked on the moon, and will shortly colonize Mars. Imagine what we could do in a million years!

The interview between Joe Rogan and Brian Cox ended with the idea that advanced civilization may not want to contact us because we are too crude, too backward. Brian Cox said that he was thinking in terms of starships, big things that made their signatures felt across the sky. Maybe, it's the reverse, where you have nanobots, tiny things which exist in the universe, being more energy efficient and less noticeable.

The main thing I got out of this interview was the possibility that we are the only intelligent life in the galaxy and that we have the responsibility to give life meaning because we are it, we are life!

*****

The themes of this whole book investigate several important questions:
- Was there a Big Bang?
- How did the universe evolve?
- Are there other universes?
- Is there life out there?
- Are there other dimensions?
- Is there a God?
- Are there several Gods?
- Are we alone?

There are no definite answers to these questions. When you go back in time to the start of the Big Bang, we just don't know what happened there for sure.

In the YouTube video: "What happened at the Big Bang", Dr. Don Lincoln says this: "The Big Bang doesn't explain the precise moment of creation. It explains things after the moment of creation."

Lincoln says that our visible universe is just a bubble in a sea of eternity, extending forever in all directions. The singularity was not nothing; it was extremely small, but not zero. The Big Bang was not an explosion. The Big bang was a stretching of space. No spot was special. Every place could be considered the center. Lincoln is honest by admitting that we just don't know what happened at the moment of inflation.

Dr. Don Lincoln concludes his video by saying, "In spite of the fact that we don't know everything about how the universe began, I'm constantly staggered by the fact that we know so much."

Sara Seager makes this valuable observation on the last page of her book, The Smallest Lights in the Universe: "What does our search for [life] say about us? It says we are curious. It says we're hopeful. It says we're capable of wonder and of wonderful things." [p.380]

39
God?

When I was a grad student in my M.A. program in English Literature, I chose for my thesis a comparison of John Milton's epic poem <u>Paradise Lost</u> with his <u>Artis Logicae</u> [The Art of Logic]. The thesis contended that Milton consciously applied many of the tactics of logic, as stated in his book, to his epic poem. My work in this was a drawn-out affair, boring and plodding. Why I took on the task, I will never know. If you ask my wife, I am the least logical person she knows! Trying to understand the war between God and Satan in epic rhyme was a real chore for me! Studying the devices in logic that applied to the epic poem was doubly boring. But I persisted.

I understood most of Milton's 17th century stilted 12 book Epic. The logical devices he superimposed upon his epic rhyme and the arguments he devised between Satan and God were skilled inventions. Richard Dawkins would laugh at this and say I got myself into a futile exercise there in my thesis, when I could have applied myself to something more useful, and perhaps, gone on for a Ph.D.

My M.A. year could have been easier and happier if I'd have chosen authors like Joseph Conrad, Mark Twain and John Steinbeck. As it was, my marks were shy of getting into a Ph.D. program. I became a reporter instead, underpaid and unappreciated. One of my M.A. compatriots said he was looking forward to a cushy university job by going on for a Ph.D. He probably had an easier life.

This confession is incidental to this investigation on God and the Universe. My marks, by the way, in my M.A. year were as follows:
1. A First in Restoration English.
2. A Second in John Milton.
3. A Third in Chaucer [not recommended for a Ph.D.]

*****

I see a clear connection in the War between God and Satan as falling into the top levels of the Kardashev Scale. The top level could be a creator civilization which seeds worlds with intelligent life and makes evolution towards intelligence possible. At the top end, they could have become purely spiritual beings, having shed their bodies. We would consider beings from a Type 4 or 5 civilization as falling into the super-human level of the Kardashev Scale. Who knows if they had a War up there among the stars with what we consider angels, and we are left with the story of that in our Old Testament?

The story of an Eden on Earth could have been true at some point where a perfect Adam and Eve could have lived for eons in Paradise until they disobeyed some Prime Directive from a higher being.

When one links such speculation to the Cosmology of the Universe one treads on sensitive territory where various religions might feel insulted. Nevertheless, what is man's intelligence for, but to question how we came about and where we are going?

Milton's God is a disappointment to me. He created legions of angels who turned against him. As an Almighty

God, he didn't see that coming? Not only that, but he also created fallible beings, and therefore, himself must be fallible by creating mistakes. Even man and woman had a fatal flaw when they succumbed to the temptation of the fruit. God didn't know they would.

If Jesus was God, become flesh to save us, then one could also argue that his crucifixion was his punishment for creating imperfect beings like us who were flawed.

I've opened a can of worms.

These thoughts are superficial thoughts which I don't even take seriously myself.

I admire atheists like Lawrence Krauss and Richard Dawkins because it takes courage not to bow to religious indoctrination and to stand alone, confident of your own intelligence, perhaps like Satan, with stubborn pride and self-assurance, holding one's head high against blind servitude.

I don't have that kind of courage. Nor do I believe those who disbelieve. Maybe, the old Catholic indoctrination is still bubbling inside me.

I don't believe in a God who sits on a cloud and looks down on us with his white beard, satisfied with his own wisdom to make righteous judgements.

I confess that I need to believe in something beyond me. When I am suffering, I appeal to some help from outside, from a divine Him, Her or Whatever notion I have of God to help me get through the pain and ultimately accept my own death. The world is so unfair.

I confess that a notion of God is possibly a delusion that helps you go through your own imminent dying. I don't care if my belief is right or wrong. I just know that there are times when I need a helping hand even if it's delusional. Who cares when you're dead? Atheist,

Agnostic, Believer, we all end up the same after the heart stops beating.

Is there a Heaven? I don't know. Is there a Hell? I don't know. The Catholic church has done a good job of fabricating images of what Hell looks like, so you better watch out! I would rather accept the notion of the absence of God as being a state of Hell, than the fiery furnace thing. But to force somebody to convert with threats, with guns or the point of a sword, now that is absolutely evil!

*****

I just don't know. I admit that I am weak. Perhaps, weaker and dumber than Richard Dawkins, and in that weakness, I need something to lean on. Maybe, that's as far as I am willing to go in a personal confession of some notion of God.

*****

This is not a book outlining my own beliefs anyway, but a summary of the research I've made into Cosmological topics with an attempt to understand the origin, the makeup and the destiny of existence.

Perhaps, thinkers like John Polkinghorne, a physicist who became an Anglican priest late in life, have felt the same push and pull about Faith as I have, and so I'm not the only one who is a searcher for a spiritual stability, since the ego and the strength of people like Dawkins does not appeal to me.

Even non-believers like Brian Greene, admit that they can understand people of Faith and respect their need for a Belief. [Youtube Joe Rogan interview: "Brian Greene Shares his surprising take on Religion and Science"]

Brian Greene recognizes the destructive force of religion throughout history, but then he also recognizes a value to a religious sensibility, "for some individuals it gives a connection to a historical lineage that's deeply valued." Joe Rogan, in his interview with Brian Greene, also says, "I feel like it gives people a scaffolding for ethics and morality and allows them some alleviation of anxiety and gives them a feeling of purpose."

Brian Greene recalls that his dad died when Greene was 23. His dad was not religious, but he wanted a religious ceremony. Brian Greene admits that it was deeply comforting to him even though he did not understand the Kaddish prayers.

"In a moment of crisis that was a very useful and comforting connection to have."

40
The Gods

Type 7, creator civilization

God is said to be outside of time and space. If he created the universe from "somewhere outside" of the universe, then the laws of physics do not apply to him. As C.S. Lewis describes it, Maleldil or God is different than the Oyarsa, the guardian spirits which are assigned to each planet. The Oyarsa are lesser gods, if you will.

Is that reality though? C.S. Lewis was writing science fiction, just prior and also during the Second World War.

According to Lawrence Krauss, Richard Dawkins and Brian Cox, this chapter which I'm about to write on "Gods" is redundant; it is a speculative chapter about God who, according to them, is probably a non-entity anyway.

Anything about God or a mystical Creator serves no purpose; it's not necessary. Atheists think of God as a useless distraction; in fact, they try not to think of Him or Her at all. But it is fun to pursue conjecture about a Supreme Being.

The renowned Christian apologist, Don Stewart, says that God created the Universe for Himself, for his own pleasure. Stewart also says that God created creation for his own glory [whatever that means]. Why would God need glory? Making the universe, however, did make known his power and wisdom. Why would he need to do that, unless he wanted adulation, maybe leave to the universe a sort of legacy of Himself. The question remains that if God was all self-sufficient, why create anything at all?

As to creating the world and mankind, theologians say that God did not need the world or need people because God has no lack. The usual Christian explanation is that God created people in his image, so that they could share in his overflowing love, grace and goodness through their relationships with the Trinity.

One would wonder whether God was bored even though he was associated in a Trinity. Maybe, he just wanted to see things grow within evolution and took pleasure in seeing that process take place.

The Trinity is hard to understand. It is a mystery. Muhammad, the founder of Islam, had trouble understanding this because Islam says God is One, as in the Hebraic religion.

God did not create "peers". He would not create his equals; otherwise, He would not be God, if there were two of them. According to myth, he did create lesser spirits, like Lucifer who fell by pride and took many angels with him to the dark side, away from God and Heaven.

God also gave mankind religion. The problem, as Mark Twain once said, was that mankind got the true religion, several of them! There are three main branches, Jewish, Christian and Islamic. The wars which ensued throughout history is evidence that religion was not a very good gift that God gave to mankind.

It is interesting that the Mormons believe in a spiritual evolution where they can evolve eventually to becoming gods themselves.

This brings us back to the Kardashev Scale where a Type 7 would be a creator civilization that goes around the universe seeding planets with intelligent life. Again, is this boredom or entertainment for higher beings, or is it an expression of their benevolent goodness?

If other planets have spirits, like the Oyarsa, guarding them, then one would ask was the Silent Planet, the Earth, the only one that had a fallen Adam and Eve? Was the Earth the only planet that needed redemption and salvation? If there is life on other planets, do they still have a pristine garden, a Paradise? Are beings on other planets "unsinned" where they live forever in a Paradisal Garden in communion with their God?

Or was Earth unique, a rare Earth phenomenon, where we were the only ones who sinned and needed a Christ to atone for our sins? Then, there would have been no need for alien life on other planets to evolve! If we on Earth are the only ones who needed salvation, then there was no need for other lifeforms on other planets to exist! If we are the only ones, what a mess we made of the universe. Also, if we were created in God's image, God sure made a mess of imposing His image on us!

I am a Christian. However, there are aspects in my religion of which I am ashamed, namely, the Holy Wars, if wars can be "Holy", and of course, the suppressions by the Inquisition and burnings at the stake. Christianity has done some not so nice things.

I refuse to sing "Onward Christian Soldiers marching as to war." From a religion that's supposed to be about love, that line is too offensive for my taste. The other thing I have trouble saying is, "to him be all honour, power and glory". Why does God need that recognition? Why does he need glory?

There are many things about the stories of Jesus I like but I have trouble with the stern God of the Old Testament.

*****

At the end of things, I may reject my old Roman Catholicism. But in my Christianity, I have to confess to requiring, not religion but a Faith in a higher Being, as a crutch to help me get through life, suffering and eventually to accept my own death. I don't know what the answers are.

41
A Swarm of Macrobes

Type 6 multidimensional

Maleldil was saddened when his supreme creation, Lucifer, gathered other angels to his side thinking he was equal to God. Why should Lucifer bow down and why obey when he had such magnificent power of his own?

The War in Heaven had changed the universe.

But more than a war, both God and Lucifer saw the cosmic struggle as a dance between good and evil, between what was visible and what was dark.

Would there be a time when the struggle would end, or would it go on forever? Would dark energy pull the cosmos apart? Ephesians 6:12 said, "For we are not fighting against flesh-and-blood enemies, but against evil rulers and authorities of the unseen world, against mighty powers in this dark world, and against evil spirits in the heavenly places." Maybe that is where dark energy and dark matter come from?

The Earth became a silent planet because its guardian angel, Lucifer, the Bent One, was cut off from communication with the other planets. The planets each had a guardian angel and they all communicated in a network with each other, except for Earth, the Silent Planet.

Some of Lucifer's followers gathered in a swarm in outer space around Earth buzzing around unhappily in their hatred of mankind. Mankind was created to make up for the mistake of the fallen angels. God intended for mankind to be placed higher than the angels because of Satan's pride. His followers, the swarm, remained patient and waited eons to topple mankind from its pedestal and to eventually destroy it. The swarm already had a cult following.

The swarm of bodiless lifeforms fed their hope for revenge. They were also known as Macrobes. Not all Macrobes were bad; some of them hovered around Mars, the planet also known as Malacandra. "The Bent One" had destroyed the outer layer of that planet. The inhabitants were forced to live in handramits, large deep cracks or canyons in the ground. The planet's guardian, an Oyarsa, recalled that everyone used to live on the surface before the days of the Bent One. Of the beasts that lived on Malacandra, the Hnakra, a vicious whale-like creature, had a mean streak and was hunted for sport by the hross. Therefore, Malacandra had been tainted in some ways by the huge war in the heavens.

Other than that, Malacandra was still a benign planet and remained an unspoiled world. Perelandra, otherwise known as Venus, also had not fallen, unlike the Earth.

It is Earth and the space around it which was the worry, where the evil Macrobes lived and schemed.

Just because Ransom, with the help of Merlin Ambrosius, was successful in defeating the cult led by Miss Hardcastle, Feverstone and Frost at Belbury, did not mean that the war against Evil was over. The swarm above the Earth in the emptiness of space was conspiring to regroup.

Their best target was always an educated group, which once infected, would spread its evil virus out and infect the weak-minded population.

Cambridge and Princeton in Massachusetts would do just as well as Belbury, maybe better, because Americans were generally considered to be a more gullible people. After all, they were young and naïve, the country being less than two centuries old.

Resurgence of That Hideous Strength:

Doctor Zener ran a Math Club outside of the local university. He spent time on research at Princeton. In his off hours, he circled the haunts of Professor Albert Einstein, hoping to enlist his membership into the privileged group which theorized that intelligent life, even spirit life existed in outer space.

Zener was infected by clever Macrobes who inhabited the higher atmosphere of Earth. They unraveled the hidden formulae in theorical math for him. He became a convert. The formulae which Zener then showed Einstein intrigued the great physicist.

"We have a group which is looking into it", he told Einstein. "Someday we hope to unify gravity and electromagnetism into one theory."

"I'm working on that too," admitted Einstein.

"Come and join our group," said Zener. "Several brains are better than one."

Einstein, though intrigued, commented, "I think I will just rely on my own brain." He flatly refused to join Zener's math club. It was not that Einstein couldn't work with others, nor share any acclaim with them; he simply liked working alone.

"It's not the sole credit I want," he said. "I am a private person who makes progress in his own time...and space, I might add."

Zener, however, started soliciting others to join his group, and the Macrobes grew stronger on the American side of the Atlantic, rebuilding what they previously had in

England. Like evil weeds, they could regrow and spring up anywhere, as long as the soil was fertile with gullible people. In America, there were plenty.

Maleldil and folks of good conscience had their work cut out for them. They had to be on guard constantly.

The Bent One told his followers, now that mankind has launched the James Webb Space Telescope, there is a good chance that the Earth may contact other intelligences. The Bent One took pleasure in quoting the Bible, John 4:35, "But I say, wake up and look around. The fields are already ripe for harvest." He encouraged his legions with these words. Battles may have been lost but the war was never over. Besides, if other intelligences were discovered in the universe and in multiverses, the harvest only got bigger for bringing other beings into the dark side.

The Bent One was confident in his smug way that if the select group at Princeton failed, similar to Belbury, then there were other fields ripe for the harvest. The war would go on and on. All it would take is one convert, one seduction to the dark side, and the war between good and evil will continue. The famous baseball player, Yogi Berra, was correct in his inimitable aphorism: "It ain't over 'til it's over."

42
Kardashev and the Bible

Obvious points have been made in <u>The Space Trilogy</u> by C.S. Lewis about the hazards of First Contact considering man's avarice for riches and plunder.

If one were an Atheist, who at least believed in higher orders of evolution, then speculations could run rampant about the miracles in the Bible. Many incidences in the Bible could be explained by the existence of supernatural beings from another dimension interfering with our history, as suggested by the Kardashev Scale.

Moses' people conceivably could cross the Red Sea by the power of "force shields" from the spirits in a hidden dimension. Could the fiery chariot that carried Elijah to heaven have been a rocket ship? Could the angel who announced that Mary would bear a child be a spirit from another dimension able to tell the future? The spirit that hovers over Jesus at his baptism like a dove could also be a being from an invisible world, falling in line with the Kardashev Scale.

*****

John Milton's notion of a God who is involved in a Cosmic battle between good angels and bad angels, in my mind, demeans the concept of a "real God". God is relegated to politics, no different than the fight between two rulers on Earth over territory. I would think that God should be above any such involvement.

When you think of a Cosmos with millions of galaxies and hundreds of thousands of lightyears, and a God who is supposed to have created all that, then a real God cannot be involved in petty clashes over power. God must be outside of creation, outside the confines of space and time. He, She or It cannot fall into any anthropomorphic notion of a God who is battling evil angels.

Maybe, the war between good and evil angels did take place in some realm of the Kardashev Scale, but that would involve a realm of lesser Gods, Gods who themselves have faults.

God's creation of Adam and Eve and the Garden of Eden is also questionable about an all-knowing God for creating fallible beings. This makes God fallible.

When God takes a human form, He makes Himself fallible, as well. If you've got an amorphous entity that's beyond time and space, an eternal creator, then you can rely in that Being for comfort, but when it's been personified, brought down to Earth in the form of Jesus, there's trouble, even though Jesus is a comforting being to whom you can trust your prayers.

Jesus let his followers believe in his godhood. He was crucified and so were his followers. Holy Wars followed throughout the ages in his name. Religion can have such devastating effects. The same applied to the history of Islam.

When we consider the Kardashev Scale, we think of a Type 7 civilization, a creator civilization, which controls time and space. To us that would come from a supernatural realm, interfering in our culture and leaving us with the seed of a divine child, a Jesus, with superpowers.

In Luke 4:30, the crowd rejected him and was going to throw him from the cliff. He disappeared somehow. "Jesus passed through the crowd and went on his way." In John 8:59, the crowd took up stones, "but Jesus hid himself, and went out of the temple, going through the midst of them, and so passed by."

"The Transfiguration" of Jesus takes place on a mountain where he becomes radiant in the presence of Moses and Elijah, while three of the apostles, Peter, James and John, look on. While Jesus was speaking, a cloud appeared from which a voice said, "This is my Son, whom I have chosen; listen to him." [Luke 9:28-3] If beings from Type 7 of the Kardashev Scale have control of time and space, then Moses and Elijah having a conversation with Jesus could very well have taken place.

"The Resurrection", as well, could be explained that way, that the Father, an advanced being on the Kardashev Scale had the power to resuscitate and resurrect Jesus. Mary Magdalene, in her grief, came upon the tomb with the stone rolled away.

> [11] Now Mary stood outside the tomb crying. As she wept, she bent over to look into the tomb [12] and saw two angels in white, seated where Jesus' body had been, one at the head and the other at the foot.
>
> [13] They asked her, "Woman, why are you crying?"
>
> "They have taken my Lord away," she said, "and I don't know where they have put him." [14] At this, she turned around and saw Jesus standing there, but she did not realize that it was Jesus. [John: 20:11-14]

"The Ascension", likewise, could be made possible by an evolved superpower from Type 7 of the Kardashev Scale. "[19] After the Lord Jesus had spoken to them, he was taken up into heaven and he sat at the right hand of God." [Mark 16:19]

In fact, all the miracles in the Bible could have been managed by an intervening super civilization of extraterrestrials, who gave us a special religion to save a self-destructive mankind.

These fantastic suppositions, I'm sure, have been posited in science fiction stories before. Look at Heinlein, <u>Stranger in a Strange Land</u>. Also, Dostoyevsky in, <u>The Idiot</u>.

When Christ communicated with his Father in prayer, might he not have been communicating with a higher being with telepathic powers, a being from a planet in another galaxy? Or when Muhammad spoke with Jibril, the angel Gabriel, that the Prophet might have gotten his dictations from a Level 7 being on the Kardashev Scale?

Skip ahead to the 19th century and Joseph Smith, the founder of Mormonism. Could his dictations have come from Moroni, the Angel, a superbeing of Type 7 on the Kardashev Scale?

It is not beyond the realm of imagination that all the great minds of human history might have been influenced by an extraterrestrial being: Buddha, Confucius, Jesus, Muhammad, Joseph Smith etc. who might have been instructed or touched by a divine spark from some Level 7 beings.

I do not know if such speculations would embrace the approval of Atheists who might just believe in a higher civilization which comes to us and interferes with us with their powers.

Such interference could be benign, but it could also be malicious, a long-lasting perverse game played by superior beings toying with our baby civilization, as entertainment or as scientific study.

C.S. Lewis describes this conflict of different civilizations on different planets, like Earth or Thulcandra ruled by the Bent One and Malacandra and Perelandra, ruled by the good Oyarsa. He mixes Christian allegories into his fabricated space story.

At the end of novel 2 about Perelandra, Lewis alludes to a final battle between good and evil, as described by King Tor. Revelation, in the New Testament, predicts the End of Days. Tor calls it a new beginning instead. The battle is necessary so that the Black Oyarsa [Satan] is blotted out. [p.332] In this way, Earth will correct its false start made by sin.

Tor also describes the Great Dance in the universe. He talks about dimension added to dimension.

The Great Dance is reminiscent of what things are made up of, according to some physicists like Brian Greene and Michio Kaku. Those atoms are subdivided into electrons and quarks and are nothing but tiny vibrating strings of energy. These vibrating strings create different particles according to the variations of the vibrations in the energy filaments. Vibrating strings are in a cosmic dance, as it were.

This is "string theory" that proposes 11 dimensions. When you think of it, Lewis who published the Perelandra fantasy back in 1943, used some felicitous phraseology that might be construed as predictive string theory that physicists revisited in the 1980s.

Lewis speaks in terms of seeing ribbons and cords and "the Great Dance". The author sees the novel's character, Ransom growing "older" in his understanding as Tor, makes things clear to him.

> He thought he saw the Great Dance. It seemed to be woven out of the intertwining undulation of many

cords or bands of light, leaping over and under one another and mutually embraced in arabesques and flower-like subtleties. [p.338]

Lewis describes Ransom's new insight in these terms: "...of a far vaster pattern in four dimensions, and that figure as the boundary of yet others in other worlds: as dimension was added to dimension..." [p.339]

The analogy of the stars doing a Great Dance goes back even further to the classical education taught in the Middle Ages where philosophers believed in the Music of the Spheres, which unifies all cultures and all creation. It is the "Musica Universalis".

The theory, in fact, originated in ancient Greece, a tenet held by Pythagoras, of how celestial bodies harmonize with each other in the universe.

C.S. Lewis confesses himself to be a Christian as he did in his book, Mere Christianity.

I stand with him, perhaps not as orthodox a Christian as he was in his era. He makes clever analogies to Christianity in his space trilogy, as with Maleldil forbidding the Lady and King [Adam and Eve] from staying overnight on the Fixed Land, and also, Tor, kneeling down and washing Ransom's bleeding heel, reminiscent of Jesus washing the feet of the apostles in subservience. [p.340]

The purpose of this book is not to get stuck on religion. Nor too much speculation! The upper civilizations on the Kardashev Scale are not proven. They are clever speculations. Certainly, a neat tool to remove the necessity of a Christian God from the Cosmos.

I separate Faith and Science. My Christianity is something that stands on its own in my heart as a support through life. New discoveries are welcome in my mind, as more and more things are discovered by technology. But I don't worry about the door to the "the God of the Gaps" being shut. Even then, I think people would still need a "God of the Needs."

43
Saint Peter Needs Help!

Type 7, a creator civilization:

Michelangelo shimmered into existence in front of "The Pearly Gates". He was taken aback somewhat when he saw Leonardo there. "What are you doing here?" he asked.

Leonardo arched his eyebrows upwards, eyebrows which were distinctly bushy in this spiritual realm. Always polite, Leonardo answered, "I'm waiting for permission to get in."

"So, why are we waiting? What's keeping us outside the Gates?" asked Michelangelo, obviously impatient.

"Maybe we've been bad boys!" commented Leonardo. "And we have to wait."

Michelangelo always had a temperamental nature, which suddenly erupted with old accusations.

"Hah! Maybe it was the fact that you were gay and liked boys!" accused Michelangelo. "Everyone in Florence talked about the red tunics you wore."

"I was proud of what I was and am!" rebuffed Leonardo. "And speaking of the closet, I heard a few things about you!"

Michelangelo took a long time to answer but finally admitted, "Well, since we're in front of the Pearly Gates, yes, I was gay too, but at least, I tried my best to follow the church's teachings. I should be rewarded for my hidden suffering."

"Maybe, that all doesn't matter to God," said Leonardo, "maybe it's other things we did which are holding us outside the Gates. When I knew you, you were such a jerk, really insulting to people."

Michelangelo pursed his lips in disdain. His anger rose and he was about to swear.

"Tut, tut," broke in Saint Peter. "We'll have none of that here."

He wanted to hear what the two men were arguing about, especially since they were at the very threshold of the Pearly Gates.

Saint Peter readjusted the clipboard he was holding. He muttered to himself, "I really must get an executive assistant."

Then in a louder voice, he said, "Look, neither of you is getting permission to get in until we settle this thing. Bickering is just not allowed in the Kingdom of Light. So, let's settle it, right here and right now."

Of course, 'right here' and 'right now' meant nothing there in front of "The Pearly Gates". This argument could potentially stretch out forever like a tedious ping-pong match until one or both of them put down their paddles.

Saint Peter eyed the two men and sighed philosophically. "Forgiveness is a long and winding road. Sometimes, it never ends."

He pointed to "The Pearly Gates". "As you can see, this is a gated community," he said, "and we just don't allow any old riffraff in."

Michelangelo was somewhat miffed. "Riffraff?" he asked.

Meanwhile, Leonardo smiled quietly at Michelangelo, and nodded, as if to say, 'if the shoe fits...'

Saint Peter jumped in before things heated up again. He explained, "in the old days, we categorically kept Unbelievers out. But recently, there's been a change in policy. Belief is not used as a measure anymore; it's Kindness."

No, Michelangelo didn't like Saint Peter's remarks, at all, or the look in Leonardo's eyes.

Michelangelo came to the Pearly Gates with a bad reputation to begin with for being unkind, a real grouch, during his lifetime. He was contentious even with his patrons, especially in Florence and Rome. According to him, you shouldn't base Heaven on just being "kind and nice."

Michelangelo had his own idea of standards in mind. How does God justify giving Leonardo divine genius when after all, he didn't even believe in the same God that the Church believed in? Besides, as everyone knew, Leonardo was gay and never felt guilty about it. If suffering alone was a criterion, as it should be, then Michelangelo should have had ready permission to get into Heaven.

Michelangelo was testy, especially since his osteoarthritis was acting up. They weren't in Heaven yet, where all the pain and suffering for those heavy of heart would one day end.

While on earth, Michelangelo had pounded away at marble all day long, day after day, liberating figures like David, divine works of art which were frozen in the marble. That obsessive chore came at a high price. His long-gnarled fingers ached. He did lighter work in painting only as a welcome relief because the paintbrush was easier to lift than a hammer and chisel. Besides, all that pounding on marble gave him a headache.

He felt that Leonardo had it easy. Everything was easy for him. Leonardo was always a pretty boy with a straight nose and a handsome face framed by tight curls, and he had a muscular physique!

Leonardo paraded his homosexuality openly in the streets of Florence. He was a well-dressed dandy and enjoyed life. On the other hand, Michelangelo was tortured by his sexual proclivity through no fault of his own. He kept "the big secret", wearing dark clothes and swearing to a life of abstinence and penance.

Leonardo acted younger than Michelangelo, even though he was 23 years older. He enjoyed designing costumes and the theatrical machinery for parades and pageants in Florence and Milan, so he could regale the royalty and princes there. He would prance around in his red and rose coloured tunics telling his underlings when to bring in the next float or to pull which wire to make a cherubic angel fly.

Admittedly the glitter and glitz of such showy pageantry won him the admiration of the art world of the High Renaissance. And because he was good at everything he did, his rival, Michelangelo, cast eyes of

Envy upon Leonardo, who eventually became the true definition of a Renaissance Man, excelling in painting, anatomy, engineering, mathematics and even in Michelangelo's own specialty, sculpture.

Oh, and before we forget, Leonardo was also an excellent musician, inventing new instruments like the keyboard operated bell. His biographer at the time, Giorgio Vasari, wrote, "he sang divinely, improvising his own accompaniment on the lyre."

The two men pursued their argument, ignoring Saint Peter, taking up their verbal fisticuffs again.

"I don't like you calling me mean to people," countered Michelangelo. "Besides, you were a real bastard between the two of us!"

Leonardo deftly swiped the remark away, politely observing, "I prefer the word 'illegitimate'!"

Leonardo's own father, Piero Fruosino di Antonio da Vinci, made a peasant girl pregnant, a tradition which was a common practice in those days, but Piero never legitimized his son. Maybe that was for the best because Leonardo was apprenticed to Verrochio to become a master painter, and thereby escaped his father's boring profession as a notary public.

Leonardo was a laid-back kind of guy. He accepted being illegitimate and being gay. What's the big deal? You are who you are, you make do, and you got on with life. In this respect, Leonardo was the better man. He had never expressed envy over Michelangelo who was born legitimately within a family of minor nobility. These things did not matter to Leonardo.

Saint Peter stepped in because Michelangelo was still in a belligerent mood. "Now boys, we don't have name-calling beyond this point." He drew a line across the doorway of "The Pearly Gates". They sparkled and the air beyond them shone and shimmered with a soothing warmth. The two artists shut up and kept their own counsel. They wondered how long they'd have to stand outside the gates, or if they'd ever get through.

44
A Universe from Nothing

Lawrence Krauss, A Universe from Nothing, Free Press, 2012.

Book Review:

Lawrence Krauss' book, A Universe from Nothing, was disappointing to me. What I wanted explained did not come into play until Ch. 9, "Nothing is Something", page 141.

At last, Lawrence Krauss would make clear to me in layman's terms how the universe could pop up out of nothing. Even after reading this chapter and the next, I still walked away dissatisfied that a universe could pop out of nothing without God.

The whole premise still falls in the realm of speculation, just as string theory and multiverses still are not proven. If Krauss had hoped to prove that you don't need God in the equation of creation, he has not wiped the slate clean on that score.

The first 8 chapters of the book were hard slogging for me. Maybe my problem was the style of writing with sentences that were pretty long, using terms that a layman is not used to. Maybe that's unavoidable because of the nature of the topic. You can't describe scientific ideas without using the scientific "parlance". I admit that in many chapters, I was not smart enough to understand what Krauss was talking about. If he intended to make it all clear to the average layman, he lost me.

My interest certainly was not there in the first 8 chapters. I wanted to get to how nothing created the universe!

Once I got into that area where Krauss explained how that phenomenon could happen, I had a few critical questions.

First, if the book was written with the agenda to disprove the necessity of God, I don't think Krauss accomplished that. Let's say before the beginning of time, there actually was nothing, no space, no time, not anything. That's what religion contends: that God made something out of absolutely nothing!

Then, if empty space is permeated by the Higgs field and if neutrinos keep popping in and out of existence, filling up the nothing, there is stuff there where actually something comes out of a hidden something. There's a "false nothing". Krauss admits to virtual particles.

Again, that's what religion contends: that God made the hidden something in empty space into a permanent something, the universe.

*****

Let's track through Lawrence Krauss' book, in the parts that interested me.

Lawrence Krauss admits that empty space has energy. [p. 91] Therefore, the nothing that physicists imagine is really not nothing!

Krauss writes that empty space exists and can store energy which has been proven by the Higgs field which permeates every corner of outer space. [p. 152] He also writes that quantum fluctuations imply that nothing always produces something, so nothing is unstable! He further admits that "virtual particles" come out of nowhere. My question is: Could they pop into existence from one of the hidden dimensions in string theory? So, nothing really is not nothing!

Krauss sees the very start of a universe as having equal parts of matter and antimatter. They would cancel out to zero if there really was nothing to begin with.

Lawrence Krauss cites Stephen Hawking who maintains:

> In quantum gravity, universes can, and indeed always will, spontaneously appear from nothing. Such universes need not be empty, but can have matter and radiation in them, as long as the total energy, including the negative energy associated with gravity, is zero. [p. 169]

However, Krauss sees this universe "establishing" a small asymmetry. The universe as we know it comes into being through the Big Bang and then rapid inflation. How did that asymmetry happen? Krauss does not really explain how something came from nothing! [p. 157] He basically suggests, let's "establish" that you have a small asymmetry. The word establish is just a more concrete way of saying, let's suggest, which is speculation!

Krauss concedes: "A definitive description of how this process could have happened in the early universe is currently lacking." [p. 159] There you go! No proof currently forthcoming!

Even when Krauss cites physicist Frank Wilczek in the 1980 article in *Scientific American*, Wilczek himself puts his argument in these terms: "one can *speculate*..." [p. 159]

As to the origin of organic life on Earth: Krauss takes the side of "abiogenesis" which states that life can arise naturally from nonlife. He says that most biochemists and molecular biologists take this view now. [p. 160] But that is basically what the Bible says: that God took clay and made a man and breathed life into him. Life was created out of nonlife.

When and how was the process accomplished where nonlife transformed into life? What was the process of this chemical change where dirt turned into a breathing, thinking being? Maybe, Adam was not a homo sapiens at the start; maybe he was a monkey, and maybe God had to intervene a second time when that monkey evolved to have self-awareness.

As a Christian, I would embrace the process of Evolution as a possibility within the power of a patient God who had lots of time at his disposal even from the time of a "primordial soup".

I agree with Lawrence Krauss that nature may be cleverer than philosophers and theologians. After all, if string theory is correct, if multiverses may very well exist, if neutrinos pop in and out of nothing, then to naively argue that "nothing out of nothing comes" is an outdated argument, lacking the facts of modern science. To argue that nothing out of nothing comes; therefore, you need God to create something, is surely a lazy way out of the argument too.

I have no qualms, as a Christian, in accepting the possibility of a multiverse. After all, an omnipotent God could do that!

I can understand why Richard Feynman is one of Lawrence Krauss' heroes. "My interest in science is to simply find out more about the world, and the more I find out, the better it is. I like to find out." [p. 177]

There is something else in which I agree with Lawrence Krauss. Philosophy and theology are ultimately incapable of addressing the questions about our existence. Krauss says what draws him is the exciting voyage of discovery.

Krauss looks at the pride of mankind, that we take ourselves as the pinnacle of Creation. I agree that this attitude is the same attitude that got Satan into trouble. If the world has another 5 billion years to go before the sun swallows us up, then it could be likely that we will again devolve into nothing.

The laws of physics simply do not care!

Anyway, Krauss entertains the possibility that something from nothing is a cyclical process. That we are headed into more disorder by the law of entropy. That someday we will recollapse inward to a point and then explode again to start creation all over again. Who knows? That can all happen as our scary destiny over 5 or 10 billion years.

*****

Lawrence Krauss says in his Epilogue, that a universe that came from nothing, without God, is invigorating. He sees our existence as even more amazing. [p.181] I can appreciate his view!

45
Out of Nothing?

YouTube: "Why does the universe exist? Jim Holt"

For two millennia, the Catholic Church, and religion in general, was not supposed to be questioned about the big questions, why we exist and why the universe exists. To do so, was sacrilege.

Even a clever person like the 17th century scientist, Gottfried Wilhelm **Leibniz**, inventor of calculus, who was asked why we have something rather than nothing, answered, it's obvious, it's because God created the world out of nothing at all. The equation was: God + nothing = the universe.

American journalist, Jim Holt points out then, who created God? He has a substitution for the blank if you take God out of the equation: Science + nothing = the universe.

He argues that physicists now have plausible arguments how the universe could have popped into existence out of sheer nothingness, "Quantum fluctuation out of the void!"

Holt mentions his friend, Lawrence Krauss, who argues the universe from nothing theory in his book, A Universe from Nothing, published in 2012.

The premise is that the laws of quantum field theory can show how out of sheer nothingness, no space, no time no matter, nothing, "a little nugget of false vacuum can fluctuate into existence, and then by the miracle of inflation, blow up into this huge and variegated cosmos." [Holt's words]

Holt points out a weakness in Krauss' argument in that Krauss looks at his laws in physics as divine commands. He says that Krauss has a "pseudo-religious" point of view about physics. Through physicists and atheists like Lawrence Krauss, atheists have taken on an impetus and clout that argues against old doctrines. In fact, Atheism has become more militant over the past hundred years gathering strength with stronger arguments and more articulate speakers.

Christopher Hitchens, Richard Dawkins and Lawrence Krauss are a strong trio of Atheists who belong to the Atheist Alliance International. They are strongly against religion because of its bellicose track record throughout history in persecuting free thought and in causing terrible wars.

C.S. Lewis in, That Hideous Strength, published in 1945, although a professed Christian, shied away from religious extremism and in fact, writes, in the name of Jesus, "I dissociate myself completely from all the organized religion that has yet been seen in the world." [p.415]

I find this ironically amusing, in that my own mother, Rosa, an uneducated woman who spoke 5 languages, said to me, "you know, I believe in God, but not in religion." By the same token, she believed that the moon landing was a hoax because God, she said, would never allow such a thing.

Ideally, this book intended to keep away from religion, and stick only to science, but that is unavoidable, since the book talks about exoplanets, and the origin of the Cosmos. Could a universe have really come from nothing? If so, did God have anything to do with it? If not, can we still have meaningful discussions about the origin of everything and where we are going? So, amidst all the speculations about exoplanets and extraterrestrial civilizations, I figured I'd add in a dash of religion in my book.

I am ashamed that God, if there is one, would fabricate a Hell for non-believers and take sides in Holy Wars. I can well understand why some of the world's foremost Atheists are upset.

Christopher Hitchens was an English intellectual who wrote some 30 books on culture, politics and literature, an avowed Atheist. Sadly, he died in 2011 at the age of 62.

Understandably, he gets upset with religions that persecute other religions. He said, we'd be better off without religion. He says that religion is man-made. He says that the God-botherers who pretend to know God's will and force that view on other people are "enemies".

Richard Dawkins and Lawrence Krauss would join Hitchens in that sentiment. Professor Dawkins wrote, The God Delusion, saying that religion is like a virus and in fact, a kind of insanity that spreads among crowds of humanity, often with violent results. His objection is not against the ordinary worshiper who sits in a pew on Sunday morning but against those fanatics who go out and kill people for their opposing beliefs.

Lawrence Krauss stoked the embers by taking God out of creation completely, writing a book about A Universe from Nothing. Professor Krauss wants to talk about the "real nothing". He sees science as making progress, unlike theology. He says that he was not attacking theological notions but celebrating our changing picture of reality.

> You could start with absolutely nothing...no particles, not even empty space...and maybe no laws governing that space and you could...without any miracles and without any need for a Creator...you could produce everything we see...

> YouTube: "Richard Dawkins and Lawrence Krauss: Something from Nothing."

Richard Dawkins agrees that you could get physics, matter, everything from nothing. Krauss admits that his area of expertise, Quantum Mechanics, defies common sense. He points out that the total energy of the universe could plausibly add up to Zero, "in spite of the fact that it's full of stuff." That may indeed by so, with or without God.

46 Replacing Religion?

YouTube: "Richard Dawkins and Lawrene Krauss: Something from Nothing."

Lawrence Krauss cites somebody who felt that "when you give up God, you give up human dignity". However, to him it's the exact opposite. Stephen Weinberg said that religion is the thing that's an assault on human dignity.

According to Krauss, the remarkable fact that we can ask these questions gives meaning to our lives.

One question that Dawkins and Krauss entertained was, "Can science replace religion?"

- Historically, religion provided explanations. Science has and is doing that.
- Science is based on evidence and is ever changing as evidence changes. Religion is based on Faith and is not supposed to be questioned.
- Comfort? Medicine and drugs are helping people. Dawkins says that science does not promise life after death. Krauss comments, "The last thing I'd ever want to be is stuck for eternity with my in-laws."
- It's eternity that's frightening. "I'd rather be under a general anesthetic."
- Morals? Dubious indoctrination from the Old, New Testaments and the Koran.
- Spirituality? They get that from the magnificence of what a space telescope can do when you see the

Milky Way. Krauss criticizes religion which says, "it's better for you to believe a fable than reality."
- Cosmology? Krauss says that the two things modern cosmology has taught us are, "that you are much more insignificant than you ever thought", and "that the future is miserable" but that should make you feel good, not bad.
- We are lucky to be alive today! "...and endowed with consciousness where we are on a random star in a random galaxy in the middle of nowhere, we were able to evolve a consciousness, live on a relatively quiescent planet".
- Science can provide consolation: once you accept reality, it's liberating.

Richard Dawkins recalled when he tried to introduce his charitable foundation, "The Richard Dawkins Foundation for Reason and Science", that he wanted young people to see where we came from and where the universe came from and be enthralled by its wonder from science and not from God.

"It's a magnificent fact that it could all have come from nothing."

He said that he applied for the Foundation to be a tax-deductible charitable organization. He pointed out that churches don't have problems doing this, but his foundation did. You have to prove that it benefits humanity. He got a letter from the British Charity Commission which asked him to kindly explain how scientific education benefits humanity. He found his having to justify his request was ridiculous.

Lawrence Kraus said that when he first got into physics, one of the biggest things he had to accept was the possibility that the laws of physics were an accident.

In another video interview, Richard Dawkins was convinced that, "Somebody as intelligent as Jesus would have been an Atheist."

*****

When we talk about extraterrestrials and religion, we have to accept the possibility that the aliens might come to Earth to force their religion upon us, like the missionaries did with primitive cultures in Hawaii, Polynesia and with the North American Indians.

However, Lawrence Krauss' book has some weak flaws. <u>A Universe from Nothing</u> still does not take God out of the equation.

For one, God is outside of creation, so he is not even within the equation. If there was a vacuity from which the universe sprang, then that vacuity was not really absolutely empty.

It was "a false nothing". Philosopher, David Albert pointed out that there was the relativistic quantum field in Krauss' view of nothing. That is not nothing.

Also, as Holt mentioned, "a little nugget of false vacuum" apparently "fluctuated into existence."

Where did this little nugget come from? It is not nothing! Even if it were absolutely nothing, who created the potential for that nothing to turn into something?

Therefore, Lawrence Krauss has not proved that God is out of the equation.

David Albert explains that Krauss misused the term nothing, since, if matter comes from relativistic quantum field, the question asks where did that field come from?

I need to point out that if indeed the universe came from absolutely nothing, isn't that what religion is all about? That God created everything out of nothing! The nothing that created the universe came from God!

47
From Another Dimension

Kate was looking for life in outer space. She wondered if her imagination was equal to the real thing. Were there terrifying gods out there, far in advance of what was on Earth? Or were there nano people that were so tiny that Earth could not detect them? Did they have warp drive despite their tiny stature?

Kate remembered seeing, <u>The Grinch Who Stole Christmas,</u> starring Jim Carey. He was a wonderful Grinch. Maybe aliens looked like him?

At the end of the movie, the camera view receded further and further away, from the town, from the mountain, further until the eye only saw the image of a snowflake. The whole story took place on a snowflake! A whole town, a whole story on the edge of a pretty snowflake!

This reminded Kate of a line in the Preface of Lawrence Krauss' book, <u>A Universe from Nothing</u>. "In fact, many laypeople as well as scientists revel in our ability to explain how snowflakes and rainbows can spontaneously appear, based on simple, elegant laws of physics." [p. xi]

Kate had not seen Jon for months. They had diverged. Kate still fed Jon leads for articles in his magazine, but they rarely met personally. E-mail was the order of the day, if Kate had a story for him.

She was immersed in her thoughts about exoplanets and about speculations of life in outer space...or in different dimensions. What tiny things were hidden within the tiny dimensions of spacetime called "calabi-yau manifolds"? There was an infinite number of them. String theory concentrated on 11 dimensions which worked. The others were inconsistent in the equations. Could these folds in spacetime hold nano civilizations like the snowflake in the Grinch story? Or could there be something bigger that was hidden by dark matter where a fantastical lifeform dwelt, camouflaged from the instruments of mankind?

Type 7 creator civilization

A proton was its own galaxy. It collided with the galaxy in another dimension and another universe was born in a Big Bang.

In that other universe, stars and galaxies coalesced and took shape. A solar system formed around one of the yellow suns. One of the planets, the third from that sun was Earth. Life evolved almost 4 billion years ago.

The proton, which was its own galaxy, continued floating in its own dimension minding its own business after its accidental collision which produced a Big Bang for a new creation.

The proton was astronomical in its own existence, being a galaxy in its own right. It was comprised of 3 quarks, which were virtual particles. They were alive with consciousness of their own. They'd always been, and they would always be. They were Alpha and Omega. They were the Father, Son and Holy Spirit, as Earthlings would come to know them. They were content in their own company but found the accidental creation of the stars in the Milky Way and other stars in other galaxies a happy accident.

Their size was immeasurable. One could think of Jonathan Swift's story of Brobdingnag where tiny Gulliver had to look up at these monstrous beasts, looking like humans but with crude manners that required entertainment like gladiatorial combat.

Thank goodness, the quarks [it might be better to call them entities] were highly civilized. They also were not like humans. They were spirits which could take human form if they wished. They moved matter around and transformed one element to another at whim. Their residence when they popped into this realm was mainly within dark energy. They were satisfied that their strength propelled galaxies apart and sent them flying outward forever in an increasing inflation of the Cosmos. The finest instruments of Earth could not detect them, nor the dark energy and matter that hid them.

They communicate sparingly with the chosen people of the Earth known as Israelites. Communication in the old days, as recorded in the Old Testament, was mainly done by the entity known as The Father. Thousands of years later, communication was taken over by the entity known as the Son, which became flesh in a Hebrew named Yeshua or Jesus. He taught mankind a gentler way than the old survival days of David and Solomon.

There was a long period in which communication with the created world remained silent. The Son lamented how he, "God-man", had been cruelly crucified at the hands of the Romans. His own people had goaded them on.

The Holy Spirit said, "These humans whom we allowed to evolve are a cruel species. Perhaps, it would be best if we snuffed them out of existence."

The other two entities voted against this notion. "Let their destiny run its full course," they said. "There is so much good, mixed in with the bad, that we cannot condemn them completely."

And so, the human race ran through its history of the Roman Empire, then the Christian Era, with Holy Wars against the Saracen and then further on through the centuries to more wars. And yet, the three spirits, in silence, noticed that there was always a spark of good in the species. The Son said, "Even one good soul would make them worthy of Salvation." The other two entities went along for the love of the One who wished to save humanity.

"I find it interesting," said the Father, "and entertaining how this species that was once an ape has evolved. They are peering at the heavens now with telescopes and they are deciphering the laws of physics. Perhaps, we will know these beings again as they get closer to finding where we exist."

"It may take a billion or two billion years," said the Son.

"We can wait," said the Father and the Holy Spirit. They were content within their own Trinity.

"We shall see," said the Son, pleased to have risen to share his rightful place once again within the unity of the Proton, joined by the Father and the Holy Spirit.

"We are Three," they said, "and yet, we are One."

They were made up of shining, sparkling filaments. The vibrations, the music they played, were preset because the speed of the vibration defined the note, and the note gave each one of them their form and identify and a harmony within the Proton. The three needed nothing but each other. They were happy to listen to the music of the spheres forever and ever.

Maybe, it was the music that made them merciful. From their perch, they looked down upon the humans they had created, and they knew the creation was good.

"We must make a place for at least half of them, if, with the grace we gave them, they have changed and won the right to be with us."

"Half may be too optimistic!" said the Holy Spirit.

"Do you remember the worthiness of the one sheep?" asked the Son.

"Indeed, I do," said the Father, "and that is why we must be merciful."

"Perhaps, we can carve out a territory in dark matter and energy for these humans when their world ends."

"Yes, carve out a Paradise for them."

"These are our children. We have watched them grow. They must now be taught."

"We do not need praise or glory from them."

"But we want their love."

A host of angels popped in out of nowhere at that moment, surrounding the trio. They sang, "They are three and yet, are One." The chant was offset in harmony by a repeat by the Trinity, "We are One and yet, we are Three."

48
Rare Earth Hypothesis

The Rare Earth Hypothesis says that things have to be just right for life to emerge. Therefore, we are special, and we are a "rare Earth".

Physicist, Brian Cox, thinks that intelligent life is so rare that it happens only once in a galaxy. He does not believe in God. He does not see that as a useful conversation.

As to the speculation of life elsewhere in the galaxy, he says, "I restrict myself to the galaxy, so I do think it's possible that at the moment there's one civilization in the Milky Way and that's us."

In a YouTube interview with Joe Rogan, Brian Cox pursues his thoughts on, "Are We the Only Intelligent Life in the Universe?" If we are, he says, that is very important because it gives us meaning, and even an ethics, on how we should behave, even politically. He sees human beings as fragile things, and if we are the only ones, then we are extremely valuable.

"There would be nowhere else where meaning exists in the Milky Way."

Cox cites his hero Richard Feynman who wrote an essay on the Value of Science.

"We are very rare configurations of atoms." Therefore, he concludes, "We are the only island of meaning in the galaxy."

Brian Cox, like fellow physicist, Lawrence Krauss, takes God out of the equation of existence. He sees the phenomenon of intelligent life as more wonderful that we exist at all, and have evolved into these special beings.

"It's more wonderful and more challenging to us because we have to take responsibility for it. We are it in this galaxy."

He does, however, believe that there is another life out there in the universe. After all, there are two trillion galaxies. "I just can't believe this hasn't happened in other places." The question to him is how often life has happened in other galaxies and how widely spaced life is.

"I think there may be one or two per galaxy on the average."

That is why humanity has to take responsibility for such a rare life. "It exists in you, and it will only exist for a short amount of time and so make the best of it."

Brian Cox entertains the idea that other civilizations may be out there, that they may be far advanced of us too because there has been so much time that's passed in the existence of the universe, 13 billion years!

"Imagine," he says, "what those civilizations would be like if they'd survived."

He points out how far we've come in only 500 years, since the time of Copernicus:
- We've gone beyond the solar system with Voyager.
- We've walked on the moon.
- We're about to go to Mars.
- Imagine what we could do in a million years!

Joe Rogan and Brian Cox mention the Fermi Paradox which asks where is the signature of all that other life? Where are they?

If we leave the evidence of spacecrafts in the solar system, then wouldn't other civilizations have left their own signatures in outer space for us to detect? Rogan poses the possibility that a higher civilization may just ignore us and Cox suggests that a nano civilization may be too tiny to detect. Then, there's the huge distance between us in intergalactic space!

*****

David Kipping is another physicist who, like Brian Cox, thinks that we may just be alone in the galaxy. Whereas Cox finds this uniqueness a thing of wonder, Kipping has a darker view of such a reality, although he also appreciates our uniqueness as special.

Kipping stresses the feeling of loneliness, "for that loneliness may be far starker than we even realize for we may very well be the only inhabited world in the universe".

The writer, Arthur C. Clarke, finds a crowded universe, as well as a single instance of intelligent life, both in fact, terrifying. The crowded universe could be filled with hostile aliens. Then, one single lifeform in a whole galaxy is also terrifying for its stark loneliness.

> A very real possibility is that we are alone and that that loneliness might be far starker than we even realize for we may very well be the only inhabited world in the universe, lost in the dark a singular candle holding back the empty void of thoughtlessness.
>
> YouTube: David Kipping in "Why we might be alone in the Universe."

However, there is one point on which both David Kipping and Brian Cox agree. And that is, if we are alone in the universe, we have a huge responsibility because we must then carry the meaning of existence.

"What a responsibility it is then to be alive. This one place, this one earth may be the diamond of the universe."

It is either a terrifying thought or a thing of wonder where you could travel for billions of light-years and all you'd ever see is countless numbers of lifeless worlds.

Kipping does share Cox's realization of wonder of our existence, if we are the only instance of intelligent life. Every one of us is incredibly special in such a universe.

Although Kipping spent some time describing the dark and lonely side of a rare Earth hypothesis, he does wax poetic, with wonder, at the end of his video by saying, "every one of the billions of people living here on this planet would be incredibly special, incredibly rare, the Diamonds of the universe."

49
Video Interview

Brian Greene: with Lex Fridman
YouTube: "Brian Greene: Quantum Gravity, the Big Bang, Aliens, Death and Meaning."

Lex Fridman: "So what's the meaning of it all?"

Brian Greene: "There has been throughout the ages some final way of articulating meaning and purpose whether it's God, whether it's love, whether it's companionship. Many people take different ways of taking this question on, and there is no one right answer, when you recognize deeply that the universe doesn't care. There is nothing out there that is the final answer. It's not as if we need a more powerful telescope and somehow if we could look deeply into the universe all will become clear. In fact, the deeper we've looked both literally and metaphorically, into the universe and into the structure of reality the more it's become clear that we are just a momentary byproduct of laws of physics that don't have any emotional content. They don't have any intrinsic sense of meaning or purpose and when you recognize that you realize that searching for the universal for this kind of question, it's a fool's errand. Every individual has the capacity to make their own meaning to set their own purpose…There is no fundamental answer; it's what you make of it."

Lex Fridman: "What is life?"

Brian Greene: "The fact of the matter is it's a continuum. There's a continuum from the things we would typically call non-living inanimate to the things that we obviously call animate and full of the currents of life…Drawing that line in the sand is not something that we're able to do. I would agree with you, it's deeply peculiar. It may in fact be unique, but it may not. It could be that the universe is such that under fairly typical conditions a star that's a well-ordered source of low entropy energy that's what the sun is, together with a planet being bathed by that low entropy energy together with a surface that has enough of the raw constituents that we recognize are fairly commonplace result of supernova explosions where a star spews forth the result of the nuclear furnace that is the core of the star. It could be that all you need is those fairly commonplace conditions and maybe life naturally forms."

Lex Fridman: "it's hard to imagine at which point this consciousness emerged…"

Brian Greene: "I think it's a continuum. I can imagine there's a lot of life out there but perhaps none of it is wondering what's the meaning of life…As a Physicist, I look at the world and see it's governed by physical laws…we got electrons, we got quarks in various flavors, we have ingredients and we have a list of laws that govern those ingredients…Is there even a hint that when you put those particles together in the right way that…an inner world should turn on…A thoughtless passionless emotionless particle when grouped together with compatriots somehow can yield something so deeply foreign to the nature of the ingredients [as consciousness]."

Lex Fridman: "I wonder if the mystery is an important component of enjoying something."

Brian Greene: cites Richard Feynman's view about people who worry that once science explains something the beauty of it goes away. Brian Greene maintains, "No, that's not the right way of thinking about it. My understanding, as a physicist, only augments my wonder. It only augments my experience…There is a wonder that comes from mystery; there's another kind of wonder that comes from deep knowing…It gives you a greater sense of awe when the curtain is pulled back…I do hold out hope that maybe before I move on to wherever [I don't think there is an after] but I would love before I leave this earth to know the answer but science and the universe is not about pleasing any individual; it is what it is."

50
The Future

Brian Greene: with Lex Fridman
YouTube: "Brian Greene: Quantum Gravity, the Big Bang, Aliens, Death and Meaning."

Brian Greene says that intelligent life on other planets would be rare, if at all. He adds, if consciousness is ubiquitous, he does not understand why aliens haven't contacted us. Presumably, he says, they should be much further ahead of us. Again, why haven't they contacted us?

If the rare Earth hypothesis is real, then we are the furthest advanced consciousness there is. We would then be in such a special place as human beings to assume such a responsibility.

However, Greene figures that we are rather run of the mill lifeforms. Other lifeforms would be far advanced of us. They would not expend the energy to hide themselves. They would not care about us.

Lex Fridman agrees with Brian Greene, explaining that humans are limited with their grasp of physics and their tools of sensing other realities. Greene adds that if we are like ants on the cosmological scale, he can imagine the super advanced aliens would not be interested, though he does not dismiss the possibility that contact may yet be coming.

Fridman suggests that the big block to sensing or communicating with aliens is the problem of language, that there is a mismatch where we miss their signals. He sees the ambition of mankind, if we are the only ones "out there", as needing to explore space, to colonize it.

Brian Greene sees it as both a physics and an engineering problem. Physics with the larger ambition of building spaceships that can travel close to the speed of light. Engineering by colonizing our solar system in a realistic way. Of even terraforming other planets like Mars.

Greene sees the fundamental part of the human spirit as one of space exploration. He sees Mars as being colonized in the near future.

However, he warns about the possibility that an advanced civilization might just destroy itself which may be a commonplace occurrence. This could be the other answer to the Fermi Paradox, why aren't they here, because they blew themselves up. They destroyed themselves.

Putting that horrific possibility aside, Greene says, once we grapple with the engineering challenges and manage them, then we can focus on physics and figure out how to travel near the speed of light.

Fridman says that when he grew up, in the Soviet Union, as a little kid, he dreamt of going to Mars, even though his field is AI [Artificial Intelligence]. Brian Greene agrees that kids need to be inspired by dreams. Fridman feels that life is too short not to do something exciting...like going to Mars. He would snap up the opportunity.

Culture seems to have inured us to the fear of death, that you don't really experience it during a career or during your youth and your life of health. It creeps up on you during old age or terminal sickness.

Fridman makes a connection that the terror of death is the creative force that advanced discoveries in science, art and literature. That your feeling of a deadline, gives you the energy to get something done. He sees his job as a robotics engineer as transferring a certain dread and feeling of mortality into his robots. He suggests that you need to feel that if you want to fit into human society, if you are a robot, you need to feel a sense of mortality. He figures that the dread of one's mortality is something that is part of the human condition. He wonders if that could be transferred to a robot, so that they'd fit in.

Lex Fridman ends his interview on an amusing note. He says, we need death to appreciate life. He adds at the end, that if robots see their own mortality, that, "it'll be us and robots drinking beers looking up at the stars, having a good laugh in awe of the whole thing."

Michio Kaku interviewed by Patrick Bet-David
YouTube: "The Future of Time Travel, Aliens and the Universe – Dr. Michio Kaku"

Michio Kaku likes to quote Galileo: "The purpose of science is to determine how the Heavens go; but the purpose of religion is to determine how to go to Heaven." He figures that you need to keep the two separate. The problem occurs when people who are in the sciences pontificate about ethics or when religious people pontificate about natural laws. As long as we keep these two things separate, they are complementary.

In the interview with Patrick Bet-David, Michio Kaku sees a future world where people's personalities will be digitized. That is a kind of immortality.

Another is biological advances where we can figure out how to stop the aging of cells. "Maybe at the age of 30, stop!" We are unraveling the question: Why do we have to die? Is there a good side to dying? Kaku answers that you don't want to stagnate. He says, "30 is a good age to live forever." He argues that with gene therapy we can reverse the process of aging, correct the mistakes in our cells.

Michio Kaku maintains that technology has a moral direction. Most scientists figure that technology is neutral, but Kaku is optimistic and gives technology a moral spin. He sees the internet as giving out information. Information is good and empowering to keep dictatorships away. The internet creates and promotes democracy.

<u>YouTube:</u> "Michio Kaku: 3 mind-blowing predictions about the future."
1. There is a new vision emerging. For Elon Musk of SpaceX, it's to create a multi-planet species. Multi-billionaire, Jeff Bezos of Amazon, wants to make Earth into a park, so that all heavy industries, all the pollution goes into outer space. Amazon would, of course, become a delivery system connecting the Earth to the Moon. Carl Sagan said that humans need to become a two-planet species as an insurance policy because some day we will be hit again by an asteroid similar to what destroyed the dinosaurs. If Elon Musk wants to settle Mars with a million colonists, the most efficient way to do this is to get self-replicating robots to do the grunt work of building shelters and support systems. Michio Kaku sees Mother Nature and the Laws of Physics as having a death warrant on humanity. Our destiny, therefore, is to be in outer space.
2. The Manhattan Project gave us the Atomic Bomb. The Genome Project allowed us to map the genome of the human body. President Obama initiated the Connectome Project to map the entire human brain. It is possible to connect the brain to a computer. What will we be able to do when was

have a direct brain-computer interface? We will have infinite knowledge. People will wear glasses where a computer is embedded in a lens and the internet is instantly available. Beyond that, we will communicate mentally. To think emails, images and memories. We are already able to record memories. We've done it up to monkeys. Someday, we will have "brain-net" which will send emotions and memories instead of zeros and ones.
3. We will defeat cancer. Physicists were innovators in every step of progress, including the industrial revolution with steam power [laws of thermodynamics], then electricity with the laws of electromagnetism, the light bulb, television and radio.
    - The transistor and the laser were invented by physicists and electrical engineers.
    - Artificial Intelligence and nanotechnology are also making headway in robotics and surgery. Biotechnology and artificial intelligence will revolutionize the job market. Nanomedicine will target cancer; individual molecules can be eradicated with precision. There will be toilets that can detect the first signs of cancer years before a tumor forms. Prof. Kaku sees cancer, someday, being no more serious than the common cold.

51
Conclusion

I was going to have one short sentence to my Conclusion: "I don't know".

But that is false humility. I do have opinions on some things that I've researched, although some of the thinking, I confess, is beyond me.

I don't want to cheat my reading audience to say at the end, I don't know, so I will fill out some final thoughts about Origins, both the Big Bang and of man, as well as the universe, multiverses and whether we are alone in the Cosmos.

It sounds reasonable that the universe began with the Big Bang, either from nothing or from a singularity. Maybe God had something to do with it; maybe not! He wasn't necessarily necessary. The notion of God, however, can be comforting to many people, including me.

I found it interesting that one of the early proponents of the Big Bang was a Belgian priest, Georges Lemaitre, who argued that the physical universe was initially a single particle, "the primeval atom". His paper on the subject came in 1927 in the Belgian language and was not noticed until 1930 when Arthur Eddington arranged for the publication of an English translation.

Most serious biologists now accept Charles Darwin's Origin of Species, published in 1859, as scientific fact. We've certainly had enough time on Earth for this to happen, 4.543 billion years of evolution.

That advanced civilizations from the Kardashev Scale could have seeded Earth and other planets with life that would evolve into intelligent life could be entertained as a possibility.

I found it more comfortable to speak of that possibility in terms of science-fiction stories of which I've had fun creating several anecdotes of imagined higher civilizations. The same applies to multiverses and other timelines involving variations of Earth, Aerth and Erth. That was fun.

I agree with Brian Cox and David Kipping that we are a rare breed, maybe the only breed, of intelligent beings in all of creation, including our galaxy and other galaxies, whether in this universe or in a multiverse. We are it! And as such, we have a responsibility in the sense that we are the ones who give meaning to existence.

This is, of course, taking the view that there is no God. If there is a God. However, if there is a God, then we also carry a sacred responsibility that gives life meaning.

I am a Christian and I've tried to keep an open mind in my research about what is or might be possible in all this science. I've kept a separation of Faith and my Understanding of the science, looking at the explanations of scientists, like Richard Dawkins, Lawrence Krauss, Brian Greene, Brian Cox and David Kipping. I've tried to make my research explainable and understandable to my readers in plain language in this book.

*****

My primary objective was to forward information and teaching in my book. Hopefully, more people will become aware of the string hypothesis, the multiverse and the speculation of aliens out there. It would be great if the general public became more interested in science. With the photos coming from the James Webb Space Telescope, more people should get excited about this new era of discoveries in existence. There is so much to learn!

Afterword

One of my favourite writers, Robert Parker, apparently died slumped over his typewriter with who knows what valuable words he had to say for his next great novel.

I think I'd like to go like that, doing what I was best at doing, even if it was over a work that was left undone like an unfinished symphony. I decided that even if I keeled over at the start of my first sentence to my new novel, I'd take a chance on starting it anyway.

I've already had the seeds planted in my mind for this novel, not knowing if I'd live long enough to finish it. It doesn't matter. Life and people will go on anyway, with or without me. The world won't be any smarter for missing my literary input.

So many artists have faced the same quandary. Da Vinci, for example, cried in his old age that he'd never finish the statue of his horse, *Il Cavallo*.

A person does not live forever, and maybe in the scheme of the whole universe, things just don't matter anyway.

In the universe, in all of existence, mankind is such a little thing. We must not be arrogant to think that we are the only ones who are blessed with creative intelligence and who feel compelled to leave literary excellence and little jewels of wisdom behind in all the darkness of space and time, as if that would be a way to show how great we are and that we deserve a measure of immortality with the evidence we've left behind.

We've littered outer space with a lot of satellites and telescopes. We can look at this as our legacy or as so much "space junk". Maybe, it won't ever be discovered or admired by an alien race for what bright people we are. Maybe that doesn't matter either.

However, we need to be struck with the awe of the millions upon billions of stars and galaxies out there where we don't even count in the swirling cosmos of existence. I like what Sara Seager says about existence and the Cosmos:

> I don't think it's an accident that there's a mirror at the heart of every large telescope. If we want to find another Earth, that means we want to find another of us. We think we're worth knowing. We want to be a light in somebody else's sky. And so long as we keep looking for each other, we will never be alone. [p.380 Seager]

Bibliography

C.S. Lewis, <u>The Space Trilogy</u>, Harper Collins, 75th Anniversary Edition in One Volume, 2013. "Out of the Silent Planet", Copyright 1938. "Perelandra", 1943. "That Hideous Strength", 1945.

Avi Loeb, <u>extraterrestrial, the first sign of life beyond Earth</u>, publ. 2021.

Brian Greene, <u>The Elegant Universe</u>, W.W. Norton, 2003.

Lawrence Krauss, <u>A Universe Out of Nothing</u>. Free Press, 2012,

Sara Seager, <u>The Smallest Lights in the Universe: A Memoir</u>, Random House, New York, 2020.

Myriad of <u>YouTubes</u>: watching people like Brian Greene, Michio Kaku, David Kipping, Brian Cox. Lawrence Krauss, Richard Dawkins, Don Lincoln, Sean Carroll, Neill de Grasse Tyson, Matt O'Dowd, also World Science Festival, FermiLabs, CoolWorlds and TED conferences.

Acknowledgements

When I was 41, I had open-heart surgery to replace a crusted aortic valve. It had been degenerating for years, I assume from the rheumatic fever I suffered in the refugee camp in Austria as a baby shortly after World War II. In 1985, the damage caught up with me. I could not walk stairs without huffing and puffing, nor walk a city block without vomiting. It looked like I would be dead within the year.

I want to thank Dr. Barwynski, a heart surgeon, at St. Boniface Hospital in Winnipeg, who gave me a St. Jude Valve, size 23, which has gifted me with an additional 37 years of life, during which time I got married and have written some 20 odd books.

Dr. Barwynski, squeezed me in on a Sunday for my heart surgery using Canada's Universal Healthcare System, which made it possible for me not to go into a debt which could have taken me years to pay off. If only the rest of the world would take care of its citizens that way then the world might be a better place.

I also want to thank Michael Kositsky for proof-reading the first ever novel I wrote in 2014, *The New Crusades*, and making excellent editorial suggestions for it.

My teachers at St. Jerome's High in Kitchener encouraged me in my academic pursuits and gave me a good classical education in the 1960s.

First of all, my English teacher, Mr. William Klos, who ignited a love of literature in me and who read poetry and excerpts with feeling. Also, Mr. Ronald Haston, who made history come alive, especially with his narrations and spirited introductions to every history class.

In terms of modern education, I need to acknowledge the computer, Google and YouTube for quick explanations of scientific things. I've tried to understand them and then explain them in layman's terms so that my reading audience could hopefully understand them. I hope I've succeeded in this book geared to an unscientific audience. It's been quite a chore to write and summarize physics, astronomy and the concepts of large ideas. I hope I've succeeded.

Thanks also to my wife, Marjorie, who likes to remain anonymous and who lets me sneak off at midnight, close the bedroom door, and fire up the computer with a cup of tea by my side. Insomnia has its good points.

Publication Contributions:
by John Hartig

1. Poem by John Hartig p. 43, "I Walked to Kenny's Grave Today", <u>Solitude: A Collection of New Canadian Poetry</u>, publ. 2009, Polar Expressions Publishing, Maple Ridge, BC.

2. Poem by John Hartig, "Songs of Innocence and Experience", <u>The Journey: A Collection of New Canadian Poetry</u>, publ. 20010, Polar Expressions Publishing, Maple Ridge, BC.

3. Short story by John Hartig, "Coffee Break", <u>Formation: New Canadian Short Stories</u>, publ. 2010, Polar Expressions Publishing, Maple Ridge, BC.

4. Short story by John Hartig, "Courage Getting Old", <u>From Across the River</u>, publ. 2011, Poetry Institute of Canada, Victoria B.C.

5. Center Page Photo Spread: <u>Our Canada – A Country for All Seasons</u>, "Spring Blossoms in Niagara-on-the-Lake," publ. 2012

6. Photo Design Ad Published: ARABELLA, Magazine Publication of Canadian Art, Architecture and Design, "Spring Awakenings 2012 Edition", Full page photo ad for *Granny's Boot Antiques* in Vineland, "Unique Folk-Art, Vibrant and Alive!" John Hartig Photos.

Books
Fiction
The New Crusades, John Hartig, second ed. 2021, first publ. 2015, by Tellwell, under my penname, Waldemar Guenter, avail. through Amazon and Ingram

The New Crusades: The Sequel, John Hartig, second ed. 2021, first publ. by Friesen Press, 2016, under my pennames of Waldemar Guenter and Alexander Kucharski, avail. through Amazon and Ingram

Duplicity, publ. Amazon, 2018, John Hartig. avail. through Amazon and Ingram

Who Killed Jean-Marie Leclair? A Baroque Murder Mystery, publ. Amazon, 2019, John Hartig. avail. through Amazon and Ingram

Love and Faith Trilogy, Books I, II, III, publ. Amazon, 2019, John Hartig. avail. through Amazon and Ingram

The Polish Cowboy, publ. Amazon, 2019, John Hartig. avail. through Amazon and Ingram

The Tipperary Kid, publ. Amazon, 2019, John Hartig. avail. through Amazon and Ingram

John's Shorts: Little Stories with Big Ideas, publ. Amazon, 2022, John Hartig. avail. through Amazon and Ingram

John's Hidden Gems: Short Story Collection, publ. Amazon, 2022, John Hartig. avail. through Amazon and Ingram

Things Have Gotta Get Better Than This, publ. Amazon, 2022, John Hartig. avail. through Amazon and Ingram

The Chosen: A Violin Story, publ. Amazon, 2022, John Hartig. avail. through Amazon and Ingram

Jonah's Journey, publ. Amazon, 2022, John Hartig. avail. through Amazon and Ingram.

The Sasquatch, publ. Amazon, John Hartig, 2023, avail. through Amazon and Ingram.

Non-Fiction

Time in a Bottle Trilogy, Books I, II, III, publ. Amazon, 2019, John Hartig. avail. through Amazon and Ingram

You Love Our Milk and Honey, Book I, II, publ. Amazon, 2020, John Hartig. avail. through Amazon and Ingram

The Second Wave: Living Through Trump and Covid, publ. Amazon, 2021, John Hartig. avail. through Amazon and Ingram.

77 Looking Back: My Sort of Diary, publ. Amazon, 2023, John Hartig, avail. through Amazon and Ingram.

Other

Can You Imagine? A children's picture book with poetry, publ. Amazon, 2019, John Hartig. avail. through Amazon and Ingram

Poetry Like Raindrops, publ. Amazon, 2019, John Hartig. avail. through Amazon and Ingram

Battle of the Violins, publ. Amazon, 2019, John Hartig. avail. through Amazon and Ingram

John's Photobook Series, Ball's Falls to Niagara Falls, publ. Amazon, 2021, John Hartig Photos. avail. through Amazon and Ingram

Louis Riel and Me, publ. Amazon, 2021, John Hartig, a historical fiction. avail. through Amazon and Ingram

Give Us Hopes and Dreams, publ. Amazon, 2021, John Hartig. avail. through Amazon and Ingram

Where Do Good Atheists Go? publ. Amazon, 2021, John Hartig. avail. through Amazon and Ingram

We Are Not Alone: Civilizations in Outer Space publ. Amazon, 2022, John Hartig. avail. through Amazon and Ingram.

The Cosmos: Origins and Aliens, publ. Amazon, 2022, John Hartig. avail. through Amazon and Ingram.

*****

John's Photobook Series

Who knows what Photobook is next? The possibilities are limited only by the imagination! avail. through Amazon and Ingram

The Bruce Trail

The Niagara Peninsula
8x11 and 8.5x8.5
the rest are 8.5x8.5

Ball's Falls

Port Dalhousie

The War of 1812

Sights in the Niagara Peninsula

Granny's Boot Antiques

Niagara-on-the-Lake

Morningstar Mill

Niagara Falls

5 Waterfalls in Niagara

Fair Havens
Christian Campground

Fair Havens
70th Anniversary

Two of my Favourite
Seasons

From the Bottom Up

- ➢ John Hartig Novels through Amazon and Ingram, google title and John Hartig
- ➢ John's Photobook Series ordered directly from Amazon and Ingram.
- ➢ Prints, any size enlargements, e-mail John directly to place an order. Pickup at the house, otherwise + shipping cost

John Hartig

John lives in Vineland, Ontario. His Photobooks and photo prints are available for home or office.

CONTACT

johnhartig.ca
Or just Google
John Hartig
Photography